Civil Engineering Basics:

Water, Wastewater, and Stormwater Conveyance

Daniel Mosher

ISBN: 1721713107

ISBN-13: 978-1721713103

DEDICATION

To my wife.

To the engineers who took the time out of their busy schedules to mentor and guide me through the beginning stages of my career.

Contents

List of Tables

List of Figures

One
Introduction

The Big Picture

Richard Feynman has long been a favorite of mine. Not because of his contributions to quantum mechanics (I can only grasp the Feynman diagrams on the most superficial of levels), but because of his ability to communicate complexity in a way that makes sense to the uninitiated. This doesn't appear to have been the result of specific effort. He explained things in this way because that is how he understood them. When he discussed heat, he would talk about the jiggling of atoms and excitedly move his hands back and forth to represent this motion. To some, heat is described by equations or memorized definitions, but to Feynman, it was the nontechnical language that best described heat. This is how I've thought of heat ever since I listened to Feynman's BBC series, "Fun to Imagine".

It is common for engineers to rely heavily on memorization. Anyone who has been supervised or trained by me will tell you how I feel about the perils of this mode of thinking. You will always have references for regulations or equations, but if you understand the underlying concepts, you will be a step ahead of your peers and you will be able to grasp the greater complexities of design much easier. If you rely on memorization, you will struggle in the long term.

As a new engineer, starting your first internship or job in civil engineering, you will quickly learn that you may not be as prepared as you had expected. I'm not suggesting that school is unnecessary or unhelpful, but it will not provide you with the basic information that you will need to design a project. Let's use the analogy of designing and building a house. Think of school as giving you the tools and knowledge for the construction of the foundation of a house, but now that you've started a job, you need the tools and knowledge to help you design and build the actual house. Hopefully, this book provides you with those tools.

The primary intent of this book is to draw a more distinct connection between general concepts, theory, and practical design and analysis. With that goal in mind, while there will be some discussion of formulas and calculations, most of the text will be dedicated to the underlying ideas.

Keep in mind that this book is *not* meant to be comprehensive, and it is *not* a textbook. You will not find every detail covered, and that is very much on purpose. Most texts related to civil engineering cover concepts, formulas, and theories that you will rarely see in a real-world design. They are inundated with information that is complex and too much to grasp at the beginning of your career. The primary reason for this is that most authors of civil engineering books are professors who have rarely strayed from academia or highly experienced engineers who don't remember what it was like at the beginning of their career.

Water Resources

Stormwater, water, and wastewater conveyance will play an important role in most of the projects you work on in civil engineering, unless you go into a specialized field like structural. For this reason, I think this is a good place to start building your base of knowledge. This is also an area where your expertise will be needed the most when coordinating projects with owners, architects, and contractors. This book only touches the surface of the knowledge base that you will build throughout your career, but hopefully it will help you establish a foundation of knowledge a little faster.

Layout of the Text

The text grew a bit longer than I had hoped, but I believe the information included is important. My goal was to be as concise as possible but provide as much information as I felt was necessary; to make this book readable, rather than just useful as a reference guide.

Chapter Two

Chapter 2 provides a foundational understanding of the underlying hydraulic concepts used in water resources engineering. It is important to have a basic understanding of these concepts before proceeding with design and analysis of stormwater, water, and wastewater systems.

While it is possible to design such systems without understanding these ideas, and many engineers do, a failure to comprehend these basic principles will lead to inaccuracies and poor design choices that could otherwise be prevented. Additionally, a firm understanding of the underlying hydraulic principles will allow you to solve complex problems that require some creativity to tackle.

Chapters Three through Seven

The Big Picture

The Big Picture sections are the most important, as they focus on the general concepts related to each design area. If you have a firm understanding of what you're doing and why you're doing it, the detailed design will be much easier.

Calculations

The Calculations sections will provide an overview of the calculations required for each design area. This will have some equations but will not go into the level of detail that most reference guides or textbooks typically do.

The full design of a system will require an understanding of the design components, design constraints, and calculations. For example, to select, and locate a pipe within a sewer system, you will need to select the material, understand the cover requirements, and determine the flow that must be carried. All of these will build on each other and I will provide basic design examples that will present a thorough understanding of the steps taken for efficient system design.

Design Elements

The Design Elements sections will provide an overview of the pieces and parts that can be used for design. This will provide information required for the selection and sizing of each element, so that you have a basic toolkit for the design process.

Design Process

The Design Process sections will provide a big picture understanding of the steps that should be taken for given project. Since the previous sections covered the calculations and selection of materials, this section will provide an understanding of the iterative process used for a system design. It's not feasible to cover every type of design, and there is not typically a step-by-step process that can be followed, but I do my best to guide you in the procedures for working your way through a typical design project. It can be said that there is no "wrong way" to work through a design, but proper iterative procedures will drastically improve efficiency and design quality.

Disclaimers

A Note on Figures

Understanding the ins and outs of construction drawings will be an important part of your initiation into the civil engineering profession. I debated whether to include a section on construction drawings or provide a construction drawing level of detail in some of the figures. After some consideration, I decided it would be best to provide only the most important details needed for an understanding of the overall concepts. Therefore, you will notice that the drawings won't be to the level of detail that you would find on the sheets of your construction plans.

Your firm will have hundreds of examples that can be utilized as references for construction drawings. Also, the annotation and styles of construction drawings will differ from firm to firm. Make sure you check with your project manager for some good reference projects when beginning a design. This way you can match the style and detail required for your firm.

A Note on Regulations

I will reference some general requirements in this text such as minimum cover requirements and minimum separation to provide you with a basic understanding of these concepts. However, each local utility provider or municipality will have their own set of regulations. Your first step in any design project should be to identify and review these regulations. For large projects, it is recommended that you meet with the reviewing agencies in person to discuss the regulations and your project.

A Note on Accuracy

I don't consider myself an expert on this subject by any stretch of the imagination. However, this book was well researched, and much of the information is based on actual experience. There will be many cases where opinions differ depending on preferred methodologies. That said, I make mistakes like anyone else, and I am always open to arguments that counter my understanding of a concept. It is very important to me that this book provides accurate information. I should also point out that this book has not been professionally edited, so I likely made some spelling, grammatical, and formatting errors as well. If you find any information that you feel is inaccurate, please don't hesitate to contact me at mosherd96601@gmail.com, and I will make the necessary changes in a future edition.

Two
Hydraulic Concepts

The Big Picture

To properly understand the design concepts in flow systems, it is important to understand the basic flow parameters and equations that are the foundation of hydraulics. It is true that for most civil engineering design projects, you will not need a thorough understanding of this information during the beginning of your career. However, this information is at the foundation of many of the equations used in civil engineering design, and it is important to have a solid understanding of these underlying concepts.

Flow Characteristics

Laminar Flow

The term laminar flow is derived from the word lamina, which references a thin sheet. In laminar flow, sheets of fluid particles smoothly slide over each other. Laminar flow is typically seen in slow moving water that contacts relatively smooth surfaces.

Turbulent Flow

Turbulent flow involves constant mixing and a generally chaotic motion of fluid particles. Turbulent flow is typically seen in fast moving fluids or fluids that flow over rough surfaces that constantly disrupt the velocity of the fluid particles. The concepts of laminar flow and turbulent flow are depicted in **Figure 2-1**.

Figure 2-1: Laminar vs. Turbulent Flow

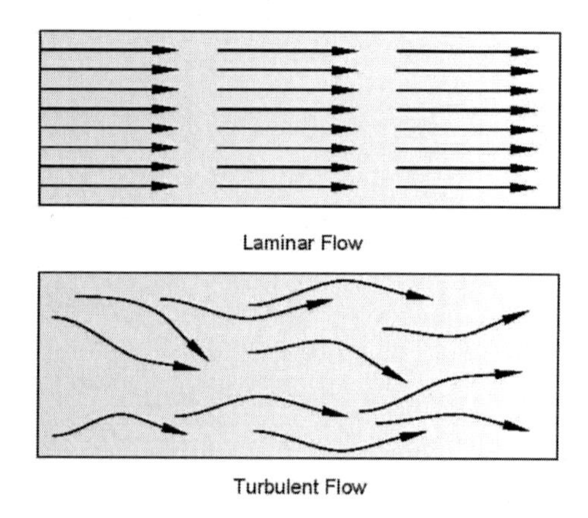

Steady vs. Unsteady Flow

Strictly speaking, steady flow is flow that remains constant at a given point over a period of time. In other words, if you pinpoint a specific particle and measure its velocity at time A and again at time B, the velocity of that particle will remain the same. If looking at the concept from a larger perspective, such as flow rate, though turbulent flow is not steady from a strict definitional standpoint, it can be assumed steady for most calculation purposes if the flowrate remains constant. For the calculations typically performed in civil engineering applications, it can be safely assumed that the flow is steady unless you are performing complex analyses over a period of time. However, these types of calculations are usually performed with software. The general concepts of steady and unsteady flow are shown in **Figure 2-2**.

Uniform vs. Non-uniform Flow

While steady and unsteady flow are concerned with the variation of the velocity of a given particle at a given *time*, uniform flow is considered flow that does not vary in *space*. In other words, if the depth and velocity at Point A and Point B are the same at a given point in time, the flow is considered uniform. If the flow characteristics vary, it is considered non-uniform flow. The concepts of uniform and non-uniform flow are depicted in **Figure 2-3**.

Figure 2-2: Steady vs. Unsteady Flow

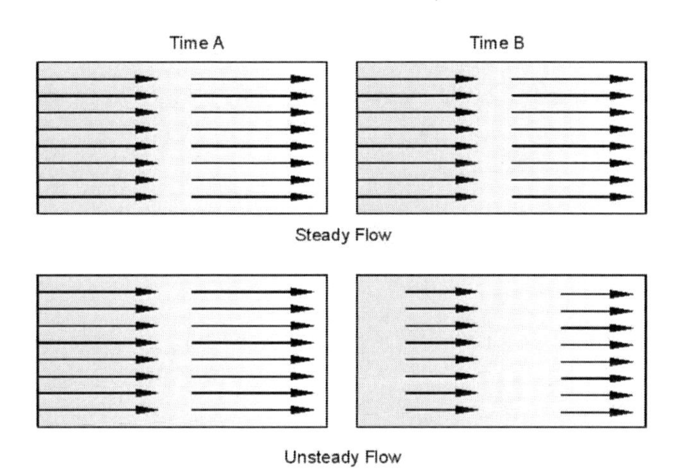

Figure 2-3: Uniform vs. Non-uniform Flow

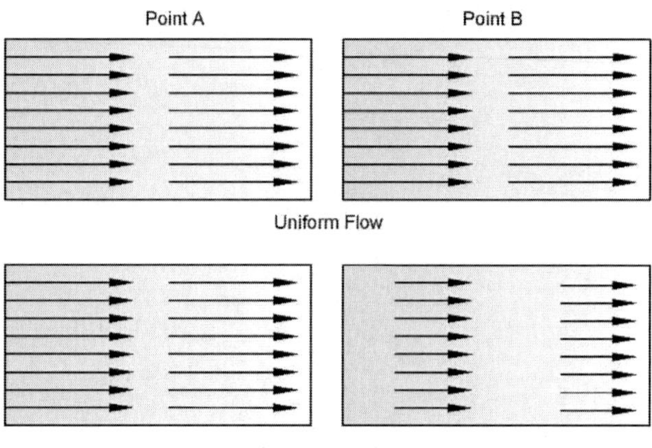

Terminology

Velocity

The velocity of water in an open channel or pipe is not the same throughout the cross-sectional area. The velocity is theoretically zero at the pipe walls and at the peak in the center of the fluid. **Figure 2-4** represents an ideal velocity profile for a full closed conduit under a laminar flow condition.[1] The variation of velocity in a conduit is greater in a laminar flow condition, whereas turbulent flow will often have a more even velocity distribution due to mixing.

Area

The area referenced in flow equations usually denotes the cross-sectional area of the fluid. Calculations for the change in area with depth of the fluid in a circular conduit can be found in **Table 2-1**.

[1] No flow in real-world settings is ideal, but this generally holds true and is used as the basis for some flow measuring techniques.

Figure 2-4: Velocity in a Pipe

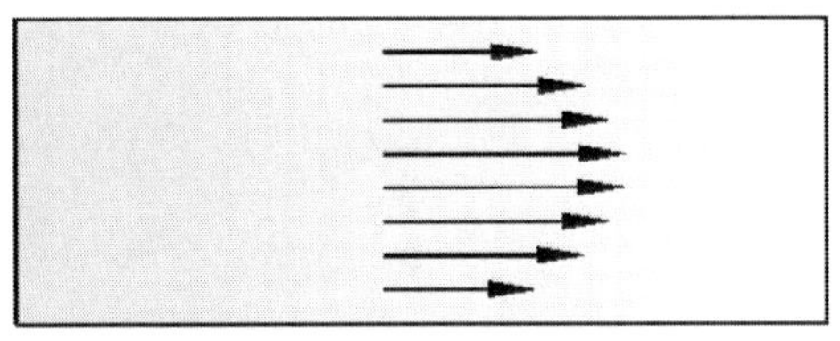

Wetted Perimeter

The wetted perimeter refers to the perimeter of the pipe or channel that is in contact with the fluid. This also varies with depth. Calculations for the change in wetted perimeter with the depth of fluid in a circular conduit can be found in **Table 2-1**.

Hydraulic Radius

The Hydraulic Radius is the ratio of the cross-sectional area to the wetted perimeter and is usually represented by the letter R:

$$R = \frac{A}{P}$$

Reynolds Number

Reynolds number can be generally described as the ratio of the inertial forces to the viscous forces in a fluid. Reynolds number can be used to characterize flow as either laminar or turbulent. The Reynolds number for circular conduits can be calculated using the following formula:

$$Re = \frac{VR_d}{v}$$

- *Re = Reynolds number (unitless)*
- *V = Average Velocity (ft/s)*
- [2]R_d = *Hydraulic Diameter (ft)* → $R_d = 4A/P$
- *v = kinematic viscosity (ft²/s)*

The characterization of the flow varies slightly depending on what source you reference, but it is generally assumed that, in circular conduits, a Reynolds number below 2,000 denotes laminar flow, and a Reynolds number above 4,000 denotes turbulent flow.

Basic Equations

Equation of Continuity

The equation of continuity shows us that flow velocity and cross-sectional flow area are directly correlated. This concept is based on the conservation of mass. The mass flow rate can be represented by the following parameters:

$$\rho AV$$

- *ρ = density*
- *A = Area*
- *V = Velocity*

Note that the units of this expression reduce as follows:

$$\frac{mass}{volume} * Area * \frac{distance}{time} \dashrightarrow \frac{mass}{time}$$

Now, following the principle of conservation of mass, we can say that the mass flow rate at point 1 must equal the mass flow rate at

[2] Note that this is hydraulic diameter, not hydraulic radius. Hydraulic radius will be used in most of the flow equations you will encounter in civil engineering.

point two in a system. This gives us the standard equation of continuity in fluid dynamics:

$$\rho_1 A_1 V_1 = \rho_2 A_2 V_2$$

Since the density will not change in a stormwater, water, or wastewater conveyance system, we can assume that the density is the same and reduce the equation to:

$$A_1 V_1 = A_2 V_2 \; or \; Q_1 = Q_2$$

The ramifications of this concept vary depending on whether you are talking about open channel flow or pressurized flow. In open channel flow, as the velocity of flow increases, the cross-sectional area of the fluid decreases. In pressurized flow, it is just the opposite; as the cross-sectional area decreases, the velocity increases. In other words, in open channel flow, the fluid dimensions can adjust, but in pressurized flow, the fluid dimensions are restricted by the pipe walls and the velocity must change to maintain the principle of conservation of mass.

Energy Equation

There are three main components of energy in fluid systems: pressure energy, kinetic energy, and potential energy. Bernoulli's equation is used to characterize the energy of an ideal fluid:

$$P_1 + \frac{1}{2}\rho v_1^2 + \rho g h_1 = P_2 + \frac{1}{2}\rho v_2^2 + \rho g h_2$$

$$or$$

$$\frac{P_1}{\gamma} + \frac{v_1^2}{2g} + h_1 = \frac{P_2}{\gamma} + \frac{v_2^2}{2g} + h_2$$

$$P = Pressure\ Energy;$$

$$\frac{1}{2}\rho v^2 = Kinetic\ Energy;$$

$$\rho gh = Potential\ Energy$$

- *P: Pressure*
- *ρ: Density*
- *v: Velocity*
- *h: height*
- *g: gravity*
- *γ: specific gravity (ρg)*

This is the most basic form of the energy equation and does not include energy lost or added to the system, a concept that will be discussed in the following sections.

Hydraulic Grade Line

The Hydraulic Grade Line (HGL) is calculated using components of the energy equation. The HGL is the sum of the pressure energy, potential energy, energy that is transferred to the system via sources such as pumps, and energy lost in the system due to friction or physical flow restrictions such as entrance or exit losses (head loss).

$$HGL = P_1 + \rho g h_1 \text{ , or}$$

$$HGL = \frac{P_1}{\gamma} + h_1 + energy\ added - energy\ lost$$

The HGL will decrease along a section of pipe due to head loss in the system. We will discuss the concept of head loss shortly. **Figure 2-5** represents a common method for measuring HGL in a system. Note that a piezometer is used, which extends from the top of the pipe. This means the velocity of the fluid will not influence the level in the piezometer.

Figure 2-6 represents an HGL profile along a section of pipe. Note the jump when the pump adds energy to the system. Also note that the slope of the HGL, which represents the rate of head loss, increases as the pipe size decreases. This is because the increased velocity (remember the continuity equation) results in a higher rate of head loss.

Energy Grade Line

The energy grade line (EGL) is the total head, which means that the velocity head is included in this calculation. Therefore, the EGL is calculated using the complete Bernoulli equation:

$$EGL = \frac{P_1}{\gamma} + \frac{v_1^2}{2g} + h_1 + energy\ added - energy\ lost$$

Figure 2-5 shows a common method for measuring EGL in a system. This is done using a pitot tube. Note that the pitot tube extends into the pipe to capture the effects of velocity within the system. Therefore, as is shown in **Figure 2-6**, as the velocity increases, the difference between the HGL and EGL also increases.

Figure 2-5: Measuring HGL and EGL

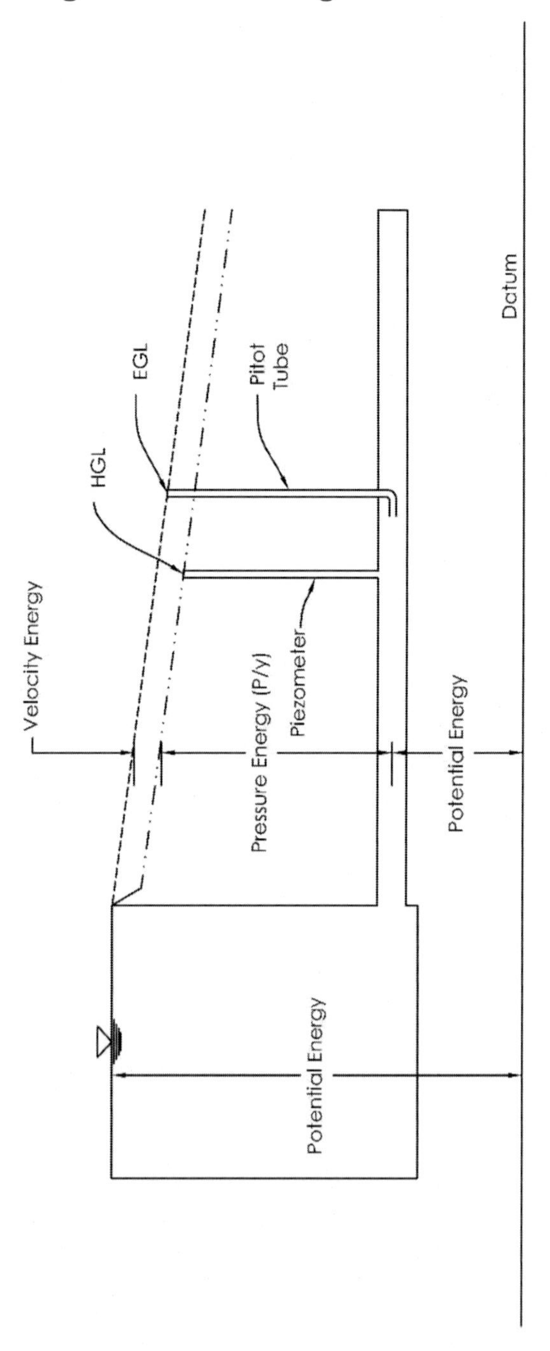

Figure 2-6: HGL and EGL in a System

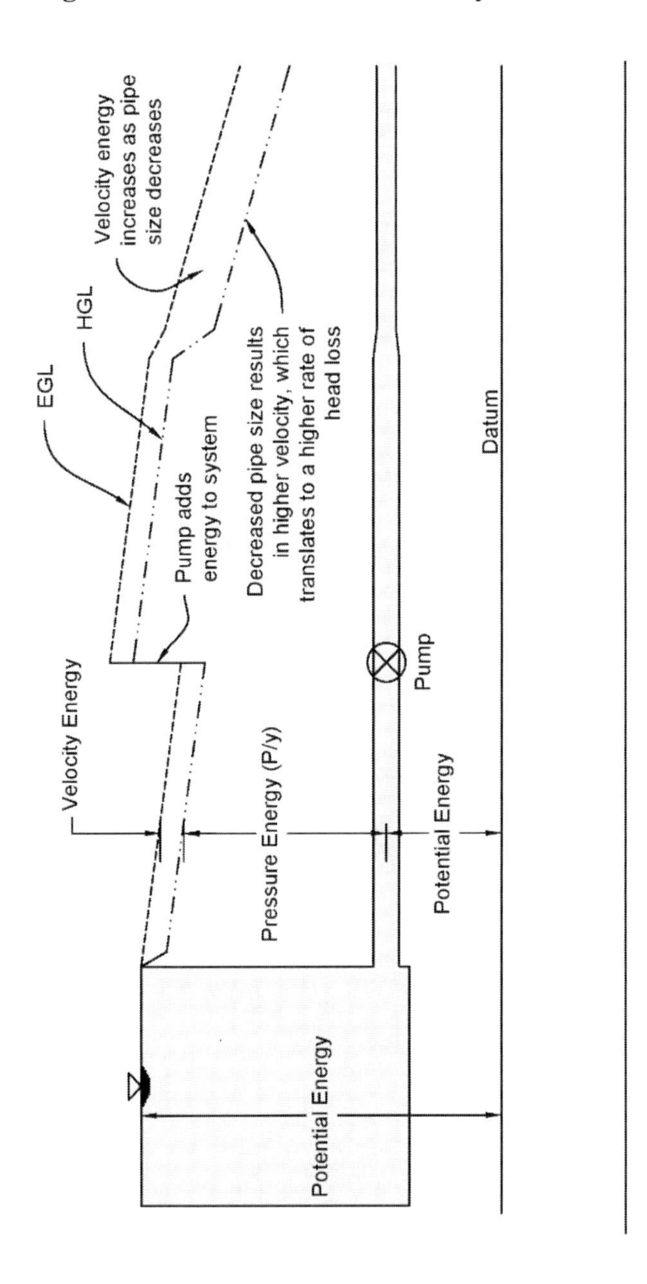

Head Loss Equations

Head loss is the term used to define the general loss of energy caused by flow resistance in a pipe network. Head losses result in a reduction in the HGL and EGL as a fluid flows through the system. This is measured in feet and is a significant factor in the design of stormwater, water, and wastewater conveyance systems. Head loss is broken down into two categories: major head losses and minor head losses.

Major Losses

Head losses termed Major Losses are those that result from friction in the pipe or channel. Unless there are partially opened valves or other major obstructions, this will account for most of the head loss in a system. Following are the standard equations used to calculate friction losses in a conveyance system.

Open-Channel Flow

Manning's Equation

Manning's equation is the most widely used and accepted equation for the calculation of friction losses in open channel flow. Remember that gravity flow in a sewer system can be assumed to be open-channel flow unless it becomes pressurized due to surcharging.

This equation is often used to determine the capacity in a system. For example, you may know that the peak loading to a pipe will be a specific number of gallons per day, so you can use Manning's equation to determine if this pipe at the designed slope can handle that flow rate. In this section, I've provided the basic equations along with a simple example to familiarize you with the use of the parameters. In future sections, I will go into further detail on how this equation is used in real-world design and analysis. **Table 2-1** provides an overview of Manning's equation, as well as equations to calculate the necessary parameters at various depths within the pipe or channel. Since this formula calculates friction losses, the conveyance material will play an important role in the resulting velocities. Manning's roughness coefficients are used to determine these friction losses. **Table 2-2** provides Manning's roughness coefficients for various materials.

Table 2-1: Manning's Equation and Parameters

Manning's Equation	
Manning's Equation for Flow Velocity (ft/s)	$v = \frac{1.49}{n} R^{\frac{2}{3}} S^{\frac{1}{2}}$
Manning's Equation for Flow Rate (cfs)	$Q = \frac{1.49}{n} R^{\frac{2}{3}} S^{\frac{1}{2}} A$
General Parameters	
n	See Table 2
Radius	$r = \frac{D}{2}$
Hydraulic Radius	$R = \frac{A}{P}$
Angle	$\theta = 2 \arccos\left(\frac{r-h}{r}\right)$
Pipe Less than Half Full	
Wetted Perimeter	$P = r\theta$
Depth of Water	$h = y$

Area	$A = \frac{r^2(\theta - sin\theta)}{2}$
Pipe More than Half Full	
Wetted Perimeter	$P = 2\pi r - r\theta$
Depth of Water	$h = 2r - y$
Area	$A = \pi r^2 - \frac{r^2(\theta - sin\theta)}{2}$

Table 2-2: Manning's Roughness Coefficients (n)[3]

Material	Manning's n Value
Asbestos cement	0.011
Asphalt	0.016
Brick	0.015
Cast-iron, new	0.012
Concrete (Cement) - finished	0.012
Copper	0.011
Corrugated metal	0.022
Earth, smooth	0.018
Earth channel - clean	0.022
Earth channel - gravelly	0.025
Earth channel - weedy	0.03
Earth channel - stony, cobbles	0.035
Galvanized iron	0.016
Gravel, firm	0.023
Masonry	0.025
Metal - corrugated	0.022

[3] (Engineering Toolbox, 2004)

Material	Manning's n Value
Natural streams - clean and straight	0.03
Natural streams - major rivers	0.035
Natural streams - sluggish with deep pools	0.04
Natural channels, very poor condition	0.06
Plastic	0.009
Polyethylene PE - Corrugated with smooth inner walls	0.009 - 0.015
Polyethylene PE - Corrugated with corrugated inner walls	0.018 - 0.025
Polyvinyl Chloride PVC - with smooth inner walls	0.009 - 0.011
Steel - smooth	0.012
Steel - New unlined	0.011

Manning's Equation Example Problem

Let's look at an example to get used to using Manning's equation. In this example, we will just use some basic parameters and develop graphs to depict the flow, velocity, and depth relationships. Let's start with the following parameters:

- *Pipe Diameter: 18 inches (1.5 feet)*
- *Pipe Radius: 0.75 feet*
- *Slope: .005 ft/ft*
- *Manning's n: 0.013*

We will need to do several calculations to create the graph, so Microsoft Excel will be the easiest way to do this. As a side note, learning excel at a high level will be very beneficial to your career. It is used frequently for civil engineering calculations. Below are sample calculations for a couple of depths to give you an idea of how the formulas from **Table 2-1** are used. **Table 2-3** shows results at each depth, and **Figures 2-7** and **2-8** show this information in graph form.

$$\boldsymbol{Example\ for\ y\ =\ 0.2ft}$$

$$h = y = 0.2\ ft$$

$$\theta = 2\arccos\left(\frac{r-h}{r}\right) = 2\arccos\left(\frac{0.75-0.2}{0.75}\right) = 1.495$$

$$A = \frac{r^2(\theta - sin\theta)}{2} = \frac{0.75^2(1.495 - \sin(1.495)}{2} = 0.140$$

$$P = r\theta = (0.75)(1.495)$$

$$R = \frac{A}{P} = \frac{0.140}{1.121} = 0.125$$

$$Q = \frac{1.49}{n}R^{\frac{2}{3}}S^{\frac{1}{2}}A = \frac{1.49}{0.013}(0.125)^{\frac{2}{3}}(0.005)^{\frac{1}{2}}(0.140) = 0.401\ cfs$$

$$v = \frac{Q}{A} = \frac{0.401}{0.140} = 2.864\ fps$$

$$\boldsymbol{Example\ for\ y\ =\ 1.2\ ft}$$

$$h = 2r - y = 2(0.75) - 1.2 = 0.3\ ft$$

$$\theta = 2\arccos\left(\frac{r-h}{r}\right) = 2\arccos\left(\frac{0.75-0.3}{0.75}\right) = 1.855$$

$$A = \pi r^2 - \frac{r^2(\theta - sin\theta)}{2} = \pi(0.75)^2\frac{0.75^2(1.855 - \sin(1.855)}{2} = 1.515$$

$$P = 2\pi r - r\theta = 2\pi(0.75) - (0.75)(1.855) = 3.319$$

$$R = \frac{A}{P} = \frac{1.515}{3.319} = 0.456$$

$$Q = \frac{1.49}{n}R^{\frac{2}{3}}S^{\frac{1}{2}}A = \frac{1.49}{0.013}(0.456)^{\frac{2}{3}}(0.005)^{\frac{1}{2}}(1.515) = 10.290\ cfs$$

$$V = \frac{Q}{A} = \frac{0.401}{0.140} = 6.794\ fps$$

Table 2-3: Manning's Equation Example

y	h	θ	A	P	R	Q	V
0.0	0.0	0.0	0.0	0.0	0.0	0.0	0.0
0.1	0.1	1.0	0.1	0.8	0.1	0.1	1.8
0.2	0.2	1.5	0.1	1.1	0.1	0.4	2.9
0.3	0.3	1.9	0.3	1.4	0.2	0.9	3.7
0.4	0.4	2.2	0.4	1.6	0.2	1.6	4.3
0.5	0.5	2.5	0.5	1.8	0.3	2.5	4.9
0.6	0.6	2.7	0.7	2.1	0.3	3.5	5.4
0.7	0.7	3.0	0.8	2.3	0.4	4.7	5.8
0.8	0.7	3.0	1.0	2.5	0.4	5.9	6.1
0.9	0.6	2.7	1.1	2.7	0.4	7.1	6.4
1.0	0.5	2.5	1.3	2.9	0.4	8.3	6.6
1.1	0.4	2.2	1.4	3.1	0.5	9.3	6.7
1.2	0.3	1.9	1.5	3.3	0.5	10.3	6.8
1.3	0.2	1.5	1.6	3.6	0.5	11.0	6.8
1.4	0.1	1.0	1.7	3.9	0.4	11.3	6.6
1.5	0.0	0.0	1.8	4.7	0.4	10.5	6.0

Figure 2-7: Manning's Equation Solution A

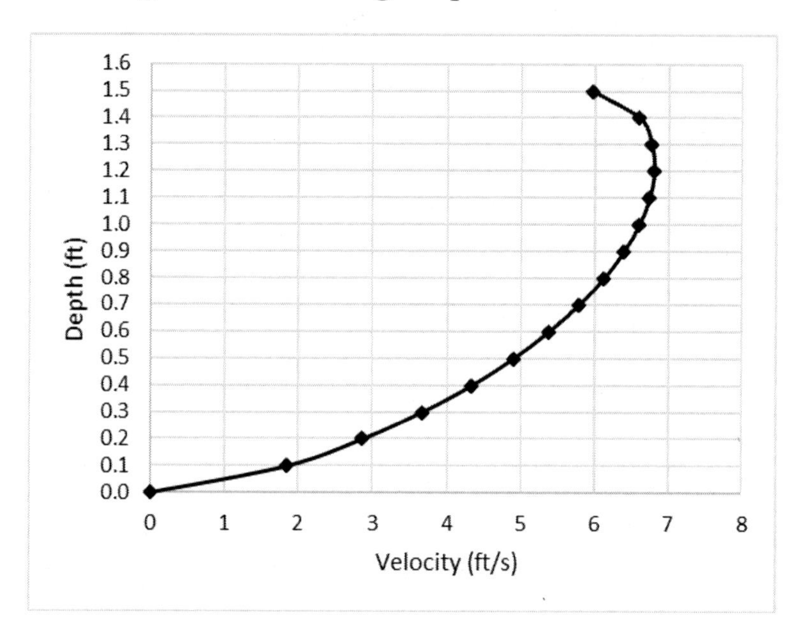

Figure 2-8: Manning's Equation Solution B

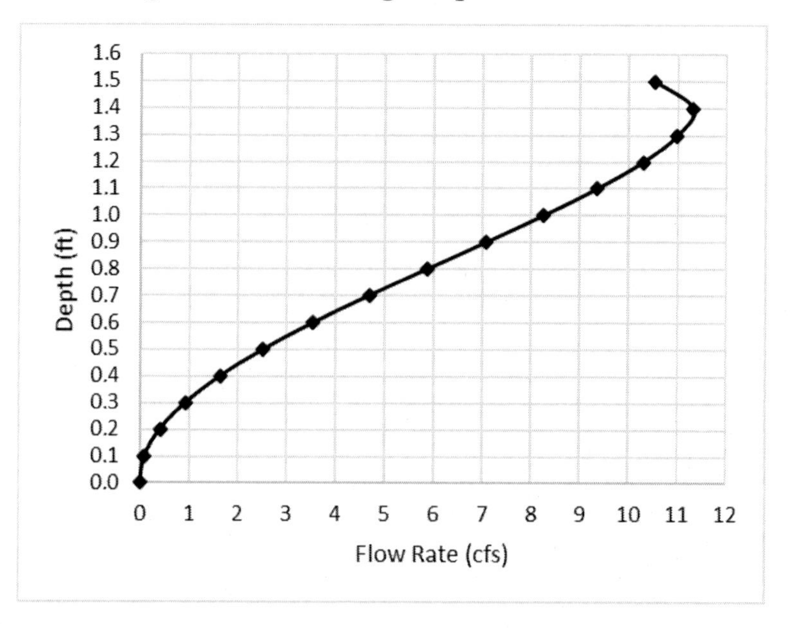

Pressurized Flow

Hazen-Williams Equation

The Hazen-Williams equation is the most commonly used for the calculation of head loss in a pressure system due to its simplicity. The Darcy-Weisbach equation is also popular but requires additional steps that are not usually conducive to a large-scale design or analysis. The Hazen-Williams equation is:

$$V = 1.318CR^{0.63}S^{0.54}$$

- *V=Velocity (ft/s)*
- *C= Hazen-Williams Coefficient*
- *R = Hydraulic Radius (ft)*
- *S = Head Loss per Length (ft/ft)*

In this formula, S is the head loss per length. Therefore, the formula can be rearranged for the calculation of head loss:

$$h_f = 10.5L\left(\frac{Q}{C}\right)^{1.85} D^{-4.87}$$

- *h_f = Head Loss*
- *L = Length (ft)*
- *Q = Flow Rate (gpm)*
- *D = Diameter (Inches)*

Table 2-4 provides a list of C-values typically used in design and analysis.

Table 2-4: Hazen-Williams Coefficients[4]

Material	Hazen-Williams Coefficient (C)
Cast-Iron - new unlined (CIP)	130
Cast-Iron 10 years old	107 - 113
Cast-Iron 20 years old	89 - 100
Cast-Iron 30 years old	75 - 90
Cast-Iron 40 years old	64-83
Cast-Iron, asphalt coated	100
Cast-Iron, cement lined	140
Cast-Iron, bituminous lined	140
Cast-Iron, sea-coated	120
Cast-Iron, wrought plain	100
Copper	130 - 140
Ductile Iron Pipe (DIP)	140
Ductile Iron, cement lined	120
Metal Pipes - Very to extremely smooth	130 - 140
Plastic	130 - 150
Polyethylene, PE, PEH	140
Polyvinyl chloride, PVC, CPVC	150
Smooth Pipes	140
Steel new unlined	140 - 150

Hazen-Williams Example Problem

Let's do a quick example to familiarize you with the equation. We will calculate the friction loss for a section of water main that is 2,000 feet of 6-inch DIP flowing at 500 gpm. First, let's list out our parameters:

[4] (Engineering Toolbox, 2004)

- *L = 2,000 feet*
- *Q = 500 gpm*
- *C = 140* [5]
- *D = 6 inches*

$$h_f = 10.5L\left(\frac{Q}{C}\right)^{1.85} D^{-4.87}$$

$$h_f = 10.5(2000)\left(\frac{500}{140}\right)^{1.85} 6^{-4.87} = \mathbf{35.9\,ft}$$

As you can see, that is a significant loss in the system.

Minor Losses

There are also minor losses in the system resulting from flow restriction at pipe contractions, bends, valves, or any other system component that may restrict flow. There will be additional discussion of this concept in the sections on water system and pressurized sewer system design and analysis.[6] Following is the formula typically used to calculate minor losses:

$$h_m = \sum k\frac{v^2}{2g}$$

- *k = Minor Loss Coefficient*
- *v = Velocity (ft/s)*
- *g = Gravity (ft/s²)*

Table 2-5 provides a list of k-values typically used in design and analysis.

---

[5] C-values in **Table 2-4** represent the materials in a new, clean condition. Municipalities will often specify lower C-values to account for future system conditions.

[6] There are minor head losses encountered in gravity systems, however, these usually have a minimal impact on flow and there is no clear methodology for calculating these losses. As will be discussed in future sections, care should be taken to ensure the smoothest possible transitions through structures such as manholes and storm drain inlets.

Table 2-5: Minor Loss Coefficients[7]

Type of Component or Fitting	Minor Loss Coefficient (k)
Tee, Flanged, Dividing Line Flow	0.2
Tee, Threaded, Dividing Line Flow	0.9
Tee, Flanged, Dividing Branched Flow	1
Tee, Threaded, Dividing Branch Flow	2
Elbow, Flanged Regular 90o	0.3
Elbow, Threaded Regular 90o	1.5
Elbow, Threaded Regular 45o	0.4
Elbow, Flanged Long Radius 90o	0.2
Elbow, Threaded Long Radius 90o	0.7
Elbow, Flanged Long Radius 45o	0.2
Return Bend, Flanged 180o	0.2
Return Bend, Threaded 180o	1.5
Globe Valve, Fully Open	10
Angle Valve, Fully Open	2
Gate Valve, Fully Open	0.15
Gate Valve, 1/4 Closed	0.26
Gate Valve, 1/2 Closed	2.1
Gate Valve, 3/4 Closed	17
Swing Check Valve, Forward Flow	2
Ball Valve, Fully Open	0.05
Ball Valve, 1/3 Closed	5.5
Ball Valve, 2/3 Closed	200
Diaphragm Valve, Open	2.3
Diaphragm Valve, Half Open	4.3
Diaphragm Valve, 1/4 Open	21
Water meter	7

[7] (Engineering Toolbox, 2004)

Minor Loss Example Problem

Let's calculate the minor losses for a section of water main that contains two 45-degree elbows (threaded regular) and one gate valve (fully opened). Water is flowing at a velocity of 4 fps. First, we will identify the minor loss coefficients:

- *45-degree elbow: 0.4*
- *Gate valve: 0.15*

So, the total minor loss coefficient is: 2 x 0.4 + 0.15 = 0.95

$$h_m = 0.95 * \frac{4^2}{2 * 32.2} = \mathbf{0.24\,ft}$$

As you can see, minor losses are minor for a reason. They are usually very small compared to friction losses. There may be some very rare cases where you will have partially opened valves that will result in significant losses in the system.

Three
Grading

The Big Picture

Grading is one of the most important skills to learn and is applicable to most civil engineering projects. This is also an area that you likely learned very little about during school. There is no one way to grade a given site, but there are some general guidelines that a good engineer should follow. The primary goal of grading is to control the flow of stormwater. You will also need to accommodate flat building pads, drivable roads and parking areas, compliance with ADA (Americans with Disabilities Act) requirements and aesthetic concerns.

You can refer to **Chapter Four** for in-depth information on hydrologic and hydraulic calculations related to stormwater runoff. This chapter will focus on design concepts required to grade a site and properly direct stormwater to desired BMP's, storm drainage structures, and outfalls.

Understanding Contours

Contours are used to represent a 3-dimensional concept in two dimensions. At first, it may be difficult to wrap your head around this representation in a meaningful way. After some time and practice, you will be able to easily visualize the topography presented in a grading plan. **Figure 3-1** provides an overview of the basic concepts related to contours.

The biggest takeaway from the figure should be that water flows perpendicular to the contours. Also remember the basic slope calculation:

$$Slope = \frac{Rise}{Run} \rightarrow \frac{Final\ Elevation - Initial\ Elevation}{Distance}$$

These two concepts will play a significant role in grading design.

Constraints

I will use the term *constraints* to describe aspects that play a role in the grading design. A common mistake for many beginners, and some seasoned professionals, is a failure to identify *all* constraints necessary to optimize the design. For the purposes of this discussion, we will divide these constraints into three categories: Functional Constraints, Financial Constraints, and Aesthetic Constraints.

Functional Constraints

Functional Constraints are established by existing site characteristics and regulations. The most obvious functional constraint is the existing topography along the grading extents. Another common functional constraint is the proposed driveway slope. Most regulations will have maximum allowable slopes for driveways that will restrict building pad elevations or road grades.

Figure 3-1: Contour Basics

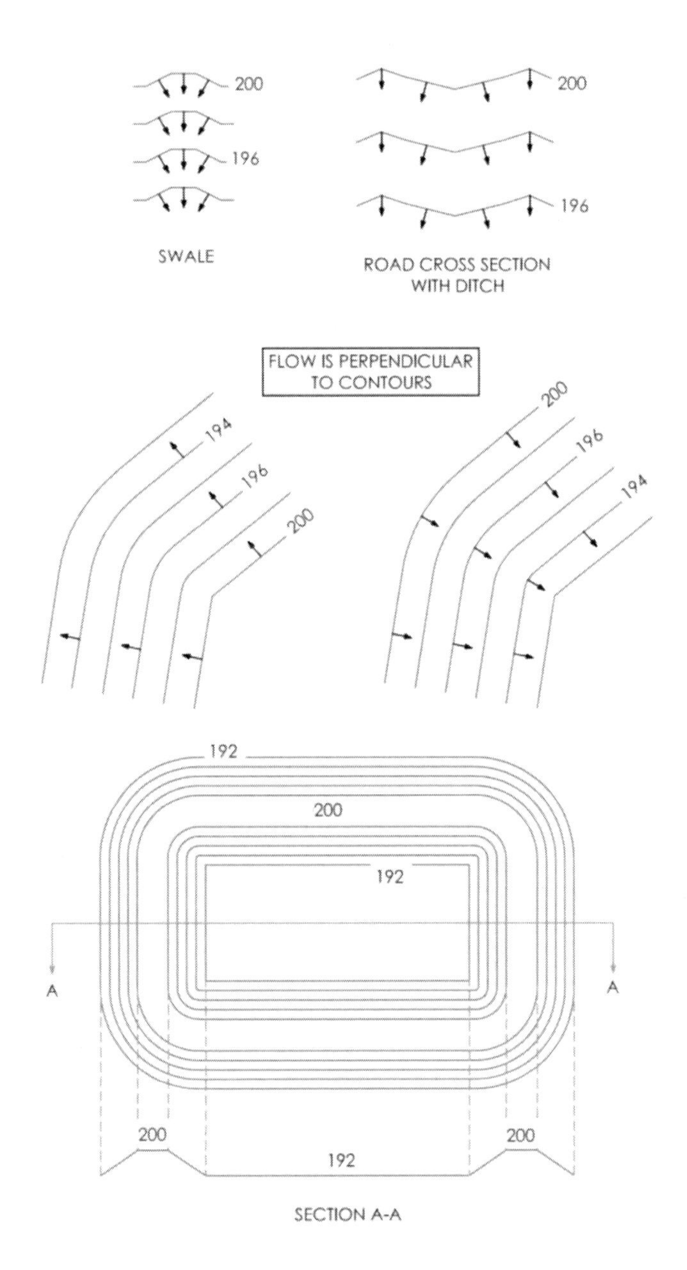

There is usually an overlap between the three types of constraints. For example, the functional constraint of the existing topography along the grading extents can be adjusted by adding a retaining wall, which encroaches on the financial constraint. You will also have the absolute functional constraint of maintaining minimum cover over a storm drainage pipe, and the financial constraint of the depth of that pipe.

Financial Constraints

Financial constraints are related to costly design aspects such as retaining walls, earthworks[8], and deep utilities. These constraints can also be impacted largely by the existing soil conditions. For example, if there is rock at relatively shallow depths, then a design that results in significant cut or deep utilities will be much costlier. You will also run into many situations where you are trading one cost for another. For example, while the installation of a retaining wall might be costly, it could also decrease the required fill. So, depending on earthworks costs, it is possible that a retaining wall could be cheaper and may be more aesthetically pleasing.

Aesthetic Constraints

Aesthetic constraints are usually determined based on discussions with the client. Before you begin designing your project, you should understand what your client wants from a visual perspective and how important this is to them. In some cases, the aesthetics, for example, the ability to see the building façade from adjacent roads is more important than the increased costs that the design may incur because of this constraint.

---

[8] The concept of earthworks and its calculation will be discussed in detail later in this chapter.

Specific Constraints

The following sections contain specific details about the most common project constraints.

Balancing Earthworks

Earthworks is the moving of dirt to adjust from an existing ground condition to a proposed grade condition. This will usually require both cut and fill. Cut is the removal of dirt, and fill is the addition of dirt. Balancing earthworks (maintaining an equality between cut and fill) is a primary goal in grading design. It is ideal to utilize the dirt on site without having to haul unused dirt away or bring new dirt to the site. The less total dirt is moved, the better. Detailed information regarding the calculation of earthworks is provided in the **Calculations** section of this chapter.

Storm Drainage Ponds

Another constraint is the location, size, and elevations of the storm drainage pond(s). Sizing will be discussed in greater detail in the stormwater management infrastructure section, however, the location and key elevations are usually determined during grading design. The elevations that will constrain your grading design are the bottom of the pond and the emergency spillway.

Bottom of Pond Elevation

The bottom of the pond will be controlled by the outfall elevation. You will need to determine what the invert of the pipe discharging water from the pond will be, as the bottom of the pond must be above this. This seems like common sense, but you'd be surprised by how far many engineers get into their design before recognizing this requirement. One exception to this rule is an infiltration pond that doesn't have an outlet pipe. Looking at the outfall elevation will allow you to determine the minimum elevation that the bottom of the pond can be set at. You should also keep in mind that the pond bottom must be sloped to allow positive drainage, which can impact the outlet elevation significantly in large ponds.

Emergency Spillway Elevation

The top of the pond dam elevation will be determined by the volume required and the emergency spillway will usually be set approximately one foot below the top of dam. Keep in mind that the stormwater must drain from the site to the pond, so the pond dam can't be raised above any portion of the site that you want to drain to the pond. It is also important to make sure that the weir of any catch basin that discharges to the pond is above the elevation of the emergency spillway. Otherwise, stormwater would overflow from the catch basins before going over the emergency spillway. This is called **short-circuiting**, and it is a common design mistake.

Storm Drainage Infrastructure

Consideration must also be taken to ensure that the depth of the storm drainage piping is optimal. If you fail to consider this issue during your grading design, you may run into a situation where you have very deep pipes or not enough cover, especially for larger pipes. For this reason, it is important to lay out a preliminary storm drainage plan and profile to check while working through your grading design.

Sanitary Sewer Infrastructure

One of the earliest considerations in a project should be the sanitary sewer infrastructure. Unless you will be utilizing a pump station and force main, you will need to consider how the gravity sewer system will be installed. You will need to identify the manhole or pump station that you are discharging to and the minimum invert for your downstream pipe to ensure that you can work upstream and serve all the houses or buildings. You also want to give consideration to pipe and manhole depths. Again, this should be preliminarily laid out so that it can be reviewed during the grading design process.

Calculations

Adjusting the grade of a site requires the movement of dirt, which is called earthworks. The main purpose for calculating earthwork quantities is to minimize costs as this is often the largest cost for site work on a project. When excess soil must be hauled off-site or additional soil must be imported to the site, the costs can increase substantially. Therefore, it is preferable to have a balanced site.

The general calculations are usually completed by a computer program, so I will not cover the manual methods for determining cut and fill values. However, there are several manual adjustments that must usually be made to ensure that the earthworks calculations accurately reflect the proposed design.

Topsoil Adjustments

Topsoil is not suitable to be used for structural fill under parking lots, roads, or buildings. It is usually stripped and stockpiled at the beginning of the grading process and can only be reused in grassed or landscaped areas. Any excess topsoil that cannot be placed on-site is hauled off. The depth of the topsoil is usually provided in the geotechnical report. This adjustment must be made when calculating earthworks, or there can be substantial costs in addition to what was expected.

Pavement and Building Pad Adjustments

The proposed grade surface reflects the top of any road, parking area, or building slab. This means there are portions of the fill requirement that will consist of concrete, base course, or asphalt. For accurate earthworks calculations, these volumes must be subtracted from the fill requirement or added to the cut requirement depending on what area of the site they are located in. Depending on the percentage of proposed buildings and pavement, this can have a significant impact on earthworks calculations.

Utilities

When utilities are installed underground, the soil is removed and then replaced and compacted, which means there may be some difference in volumes. It is also possible that the in-situ soil is not suitable for backfill or that a large pipe is taking up considerable volume. It is typically impractical and unnecessary to adjust for these parameters as they are relatively minor, but there are some cases where this will be important, especially in situations where utilities will be very deep or if large pipes will be installed.

Adjustment Factors

The volume that a quantity of soil will take up will vary depending on the level of compaction. Correction factors are needed if soil is removed from an in-situ condition and heavily compacted for use as structural fill. Below is the general terminology for these conditions and the adjustment factors that are typically used. **Table 3-1** provides some typical adjustment factor values. Often, the adjustment factors will depend on both the soil type and experience in the region.

- BCY (Bank Cubic Yards): The volume of soil to be removed from the in-situ location.

- LCY (Loose Cubic Yards): The volume of soil after it has been removed when it is stockpiled or being transported. This will typically result in an increase in the overall volume and is often referred to as swell.

- CCY (Compacted Cubic Yards): The volume of soil after it has been compacted as fill. This will usually result in a decrease in the overall volume and is often referred to as shrinkage.

Table 3-1: Typical Adjustment Factors

	BCY	LCY	CCY
Clay	1.0	1.27	0.90
Common Earth	1.0	1.25	0.90
Rock (blasted)	1.0	1.50	1.30
Sand	1.0	1.12	0.95

To understand the concept of the adjustment factors, let's look at a brief example. Assume that you have a situation where you are cutting 1,000 cubic yards of sandy soil, transporting it to another site, and placing and compacting it there. **Table 3-2** shows the changing volumes that would result. In the future we will go through a more thorough earthwork calculation example.

Table 3-2: Calculated Soil Volumes

	In-Situ (BCY)	Transported (LCY)	Compacted (CCY)
Sandy Soil	1.0	1.27	0.90
	1,000 CY	1,270 CY	900 CY

Balancing Techniques

When grading a site, it is optimal to move as little dirt as possible. You will want to balance the site, meaning no soil needs to be hauled off-site and no soil must be hauled to the site, if possible. It is also important to understand what adjustments to make to achieve a balanced site. For example, if you have excess soil (cut), it is important to try to adjust the grading to decrease cut rather than increasing fill. Increasing fill areas in this scenario would increase the movement of more dirt around the site whereas decreasing cut areas will reduce the total movement of dirt.

Earthworks Calculation Example

Following is a basic earthworks calculation example covering several of the previously discussed concepts. Assume we are given a project that has the following parameters:

Given Cut	**Required Fill**
10,000 CY	9,000 CY
Pavement in Cut Section	**Pavement in Fill Section**
1,000 CY	1,500 CY
Topsoil in Cut Section	**Topsoil in Fill Section**
2,000 CY	2,500 CY
BCY Adjustment Factor	**CCY Adjustment Factor**
1.0	0.90

Following is an overview of the cut/fill calculations. I've laid out the formulas to show when you will add or subtract the pavement and topsoil. For example, in cut areas you can assume the topsoil has already been stripped and removed as part of the clearing and grubbing of the site. Therefore, it can be subtracted from the overall cut requirement.[9]

[9] Topsoil may be placed on-site in landscaped areas depending on the project. Make sure you discuss this with your project manager or client before making the topsoil adjustments.

$$Cut = Given + Pavement - Topsoil$$

$$Cut = 10{,}000 + 1{,}000 - 2{,}000 = 9{,}000 * 1.0 = 9{,}000\ CY$$

$$Fill = Required - Pavement + Topsoil$$

$$Fill = 9{,}000 - 1{,}500 + 2{,}500 = 10{,}000 * 0.90 = 9{,}000\ CY$$

$$Cut = Fill \rightarrow Balance\ Site$$

Design Elements

Spot Elevations

Spot elevations are marked elevations at a given point. Most grading designs will be based primarily off spot elevations at key locations. The grading design examples discussed shortly will show how this works.

Contours

A contour is a line that connects areas of the same elevation. As mentioned previously, the flow is perpendicular to the contour. The spacing of contours represents the steepness of the slope in that area.

Design Process

The most common question I get when trying to teach grading is "where do I start"? The previous sections were meant to provide an overview of the many variables that you must consider when grading a site, as it is common for even experienced engineers to neglect one or more of these factors. The most important concept to understand regarding grading is that it is an iterative process. What I mean by this is that there will rarely be a step by step progression that takes you through the grading process without returning to earlier steps and making adjustments. This is why grading can be so challenging for some engineers. You may find a design that optimizes earthworks but is not aesthetically pleasing to the client. Or you found the most aesthetically pleasing design, but the routing of stormwater will be much more challenging and costlier. Plus, there is not always one optimal answer. There can often be multiple ways to grade a site, that will each be just as optimal as the other.

In the following sections, I will take you through the process of grading a site, so you can have some understanding of how to get started when you are handed a project to work on. Although I stated previously that you cannot simply step your way through a grading plan, I will provide some guidance in a step by step format and explain the concept of iteration as I discuss these procedures.

Identify Grading Extents

It is common to just identify the property line as the grading extents, but this is not always the correct way to approach a grading design. There are several factors that may limit the extents further.

First, you will need to determine if there are any buffers that prohibit disturbance. Landscape buffers will be the most common limitation. In many cases, you can grade within the landscape buffers, but not develop within them. Even if grading is allowed, some consideration should be given to the landscaping that will be needed in these areas. You should coordinate with a landscape architect to determine what slopes will be acceptable.

You should also consider the practicality of the contractor being able to grade right up to the property line without disturbing the adjacent properties, or the existence of any other buffers (i.e. wetlands) in the area. Additionally, if silt fence is to be installed (see **Chapter 5**) then you will need to make sure the contractor has room to install the silt fence and leave it undisturbed during grading. This means you will not be able to design grading up to the property line in these areas. **Figure 3-2** shows how the grading extents might be determined on a project.

Identify Design Slopes

It is helpful to determine what your acceptable minimum and maximum slopes will be for the various areas you are grading. Here are some examples:

Table 3-3: General Slope Parameters

Land Use	Maximum Slope	Minimum Slope
Grassed Area	3:1 (33%)	2%
Parking Lot	4%	1%
Driveway	8%	1%

In most cases, there will be some leeway in these parameters. For example, you may be able to grade at a 2:1 slope in some grassed areas, as long as you have a clear understanding of the additional requirements that may be needed to prevent erosion of the slopes and the added difficulty in maintaining them. You may also be able to adjust the maximum slope in a parking area if the conditions make it impractical to maintain a maximum of 4%. The maximum slope of 4% is usually based on preference rather than regulation and can therefore be increased. These types of changes should be discussed with your client.

Figure 3-2: Identifying Grading Extents

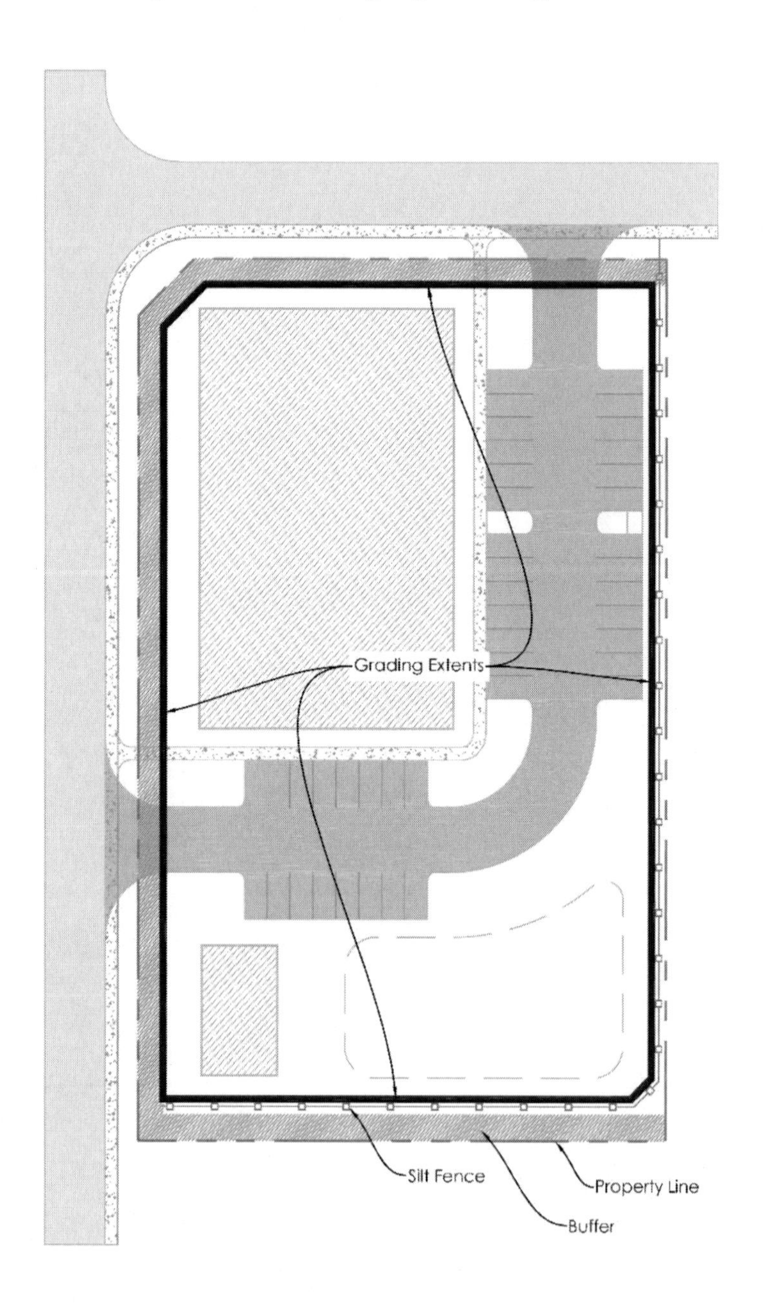

In some cases, the minimum and maximum slopes will be set in stone. For example, if you are connecting a commercial driveway to a DOT road, you will have maximum changes in elevation that are allowed. Once you get more comfortable with grading, you won't need to list out these design parameters, but when you are starting out, I recommend listing these out and discussing them with your project manager before you get started.

Identify Design Preferences

Aesthetics will be an important component of the grading plan for most clients that you deal with. For example, if you are designing a commercial business, they may want the building to have visibility from the adjacent roads. If possible, you should also consult a landscape architect to discuss aesthetics.

Analyze Tie-in Elevations

The next step will be to review the existing topography and identify the key tie-in elevations that may restrict your design. These are the elevations around your grading extents. Key tie-in elevations will be at driveways and any areas where your proposed parking area, building, or roadways will be close to the grading extents.

Identify Key Constraints

We've already discussed the constraints involved in grading. This is the step where you want to review those constraints and identify the constraints that will have control over your design. In some cases, the invert of the manhole you are tying the sewer into may elicit some control over your design. In others, the pond outfall may significantly control your design. In this stage, you will need to start getting a rough idea of how your stormwater and gravity sewer will be routed. You will also need to begin preliminary storm drainage calculations to start laying out your stormwater management design. **Figure 3-3** provides an overview of some constraints that may be identified during the grading process.

Set Key Elevations

During this step, you will start setting some key elevations. These may be building pad elevations, top and bottom of pond elevations, roadway elevations, etc. Make sure you start with the big picture and work your way into the details. It is tempting, especially with programs that allow you to view a 3D representation of your grading, to start setting bottom and top of curb elevations or filling in all the elevations for parking islands. However, this will cost you a significant amount of time in the long run. First, get your general design set up with all the constraints worked out, and then work your way into the finer details.

Begin Earthworks Analysis

There is a very important reason that I've specified the need to identify key constraints prior to looking at the earthworks. I have seen several cases where a site is fully graded, and the earthworks is perfectly balanced, only to find that the other constraints were not considered, and the design must be completely redone. It is always preferred that earthworks are properly balanced if possible, but the previously discussed constraints will usually govern.

Figure 3-3: Identify Key Grading Constraints

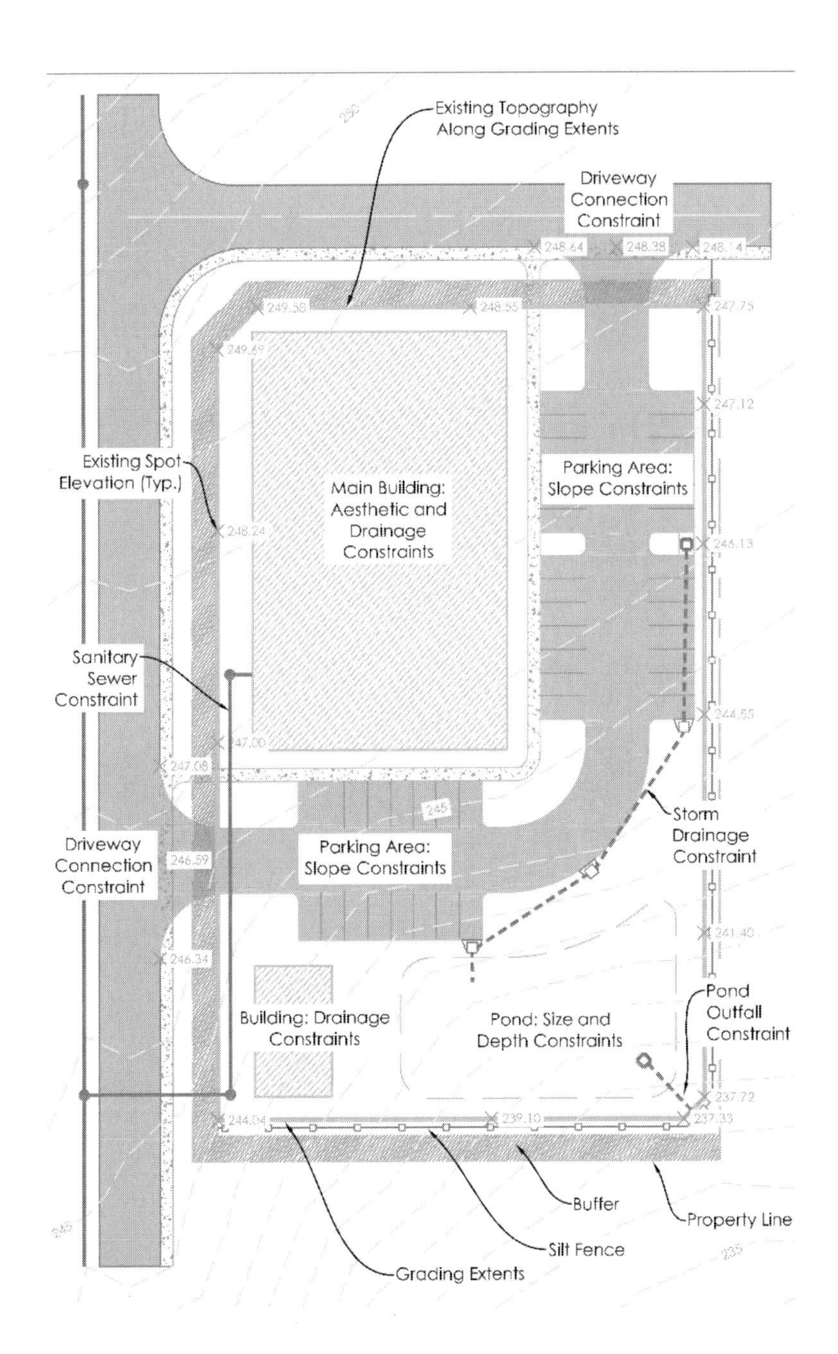

Repeat

You will rarely get it right the first time. You will need to go back and make changes based on the constraints to come up with the most optimal design. The good thing is that at this point, you will have a solid understanding of your constraints and how they will impact your grading design, so you can work toward the optimal solution methodically rather than just guessing.

There is no way to cover every grading scenario with examples. Grading will take time and experience, but hopefully this information will help you get started and learn faster. **Figure 3-4** illustrates an example of what a grading design might look like. As I stated in the introduction, this is not to a construction plan level of detail, but the concept should be clear.

Figure 3-4: Grading Example

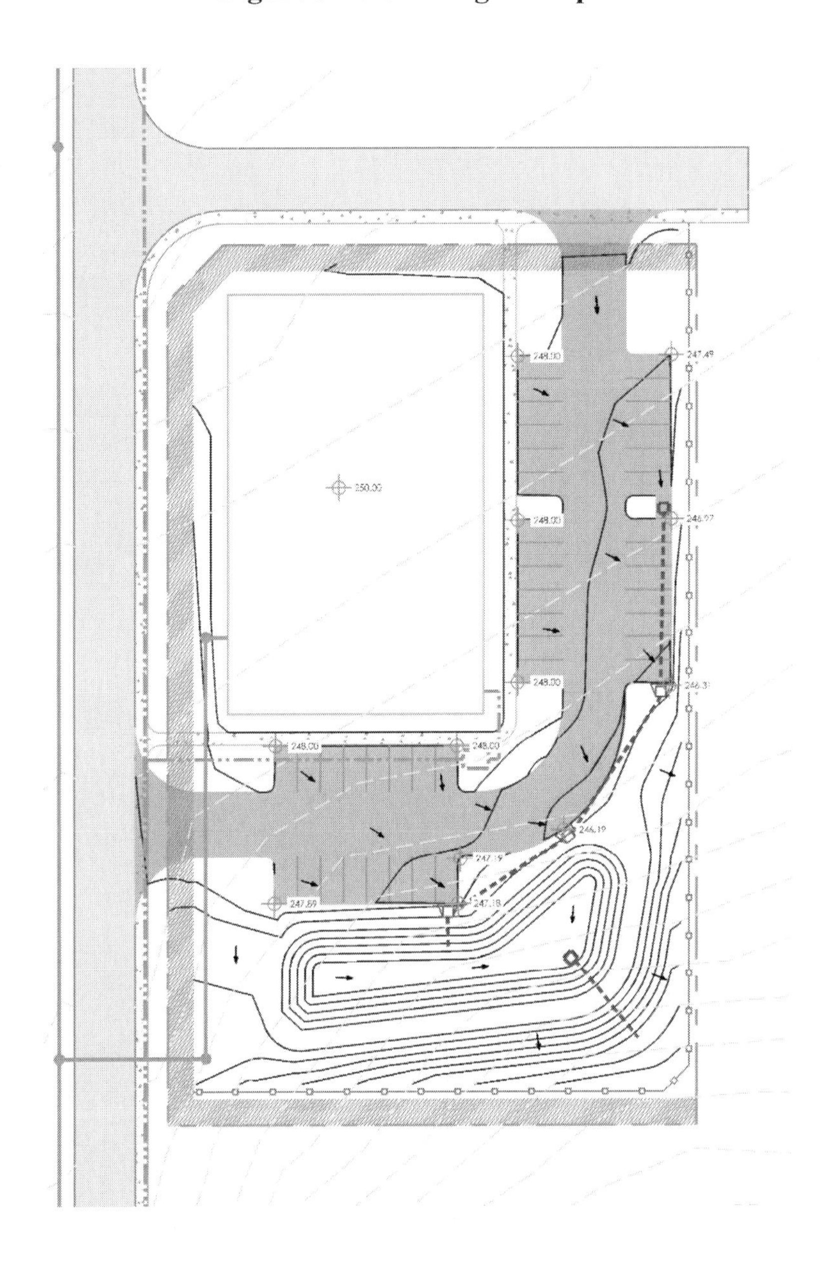

Four
Stormwater

The Big Picture

In 1950, it was estimated that the world's urban population consisted of 746 million people. By 2014, this number increased to approximately 3.9 billion, and 2.5 billion will likely be added by 2050 (United Nations, 2014). This continued increase in urban populations and the subsequent increase in urban surfaces has made stormwater management a key environmental concern.

There are two main components of stormwater management: quantity control and quality control. Quantity control consists of the management of stormwater (higher peak flows and added volumes) that will result from the additional impervious surfaces. Quality control consists of the removal of various pollutants that may be carried by the stormwater.

Stormwater

Every time we construct a road or building, we are altering the natural process with which rainfall interacts with that altered environment. Chapter 4 will cover how we analyze this situation, but the overall concept is not complex. Consider a common example; you are designing a building and parking lot on a 5-acre parcel that is currently heavily wooded. As you can imagine, when rain falls on a heavily wooded area, a significant amount is initially abstracted by trees and underbrush. And unless the slope is very steep, the stormwater that does run off the site will take a significant amount of time to reach the outfall. However, after the building and parking lot are constructed, the runoff characteristics will change significantly. Much more stormwater will run off the site much faster. Additionally, this stormwater will gather pollutants from these surfaces as it travels to the outfall. For a long time, little was done to mitigate the impact that development had on these runoff characteristics. However, regulations are finally catching up with the environmental impacts, requiring proper stormwater management techniques to ensure sustainability as development continues to increase.

When you are designing a site, your job is to understand and ensure that all stormwater is properly conveyed to the outfall, usually through a conveyance system of pipes, structures, and swales which carry it to stormwater management structures such as detention ponds that facilitate quality and quantity control before the stormwater discharges from the site.

Quantity Control

Most regulations will require that the post-development peak flow rates are less than or equal to the pre-development peak flow rates for certain design storms. These regulations will vary with location and should be verified prior to design. Some jurisdictions will also have volume control requirements, meaning the post-developed runoff volume to the outfall must be less than or equal to the pre-development runoff volume.

Terminology (Quantity Control)

Watershed

A watershed is the area contributing stormwater runoff to a specified outfall location. Project sites can have multiple watersheds depending on the topography and proposed drainage system. When the impervious area of a project site is increased, the resulting modifications to the watershed will produce increased runoff. In addition to the increased volume of water discharging from the site, the speed with which it reaches the outfall location will also increase, generating a higher peak flow rate.

Several factors will come into play when delineating a watershed. The most obvious is the site topography. It is also important to consider other developed areas such as buildings, parking areas, roadways, and associated storm drainage infrastructure. **Figure 4-1** represents a delineated watershed. Note how the concept of flow traveling perpendicular to the contours guides the delineation. Also note the overall concept; the delineated area is the area contributing stormwater runoff to the selected outfall.

Time of Concentration

Time of concentration is defined as the amount of time it takes for water to travel from the most hydraulically distant location of a watershed to the outfall point. It is important to understand that the most hydraulically distant location does not always correlate with the point at which water must travel the furthest to reach the outfall location.

Figure 4-1: Watershed

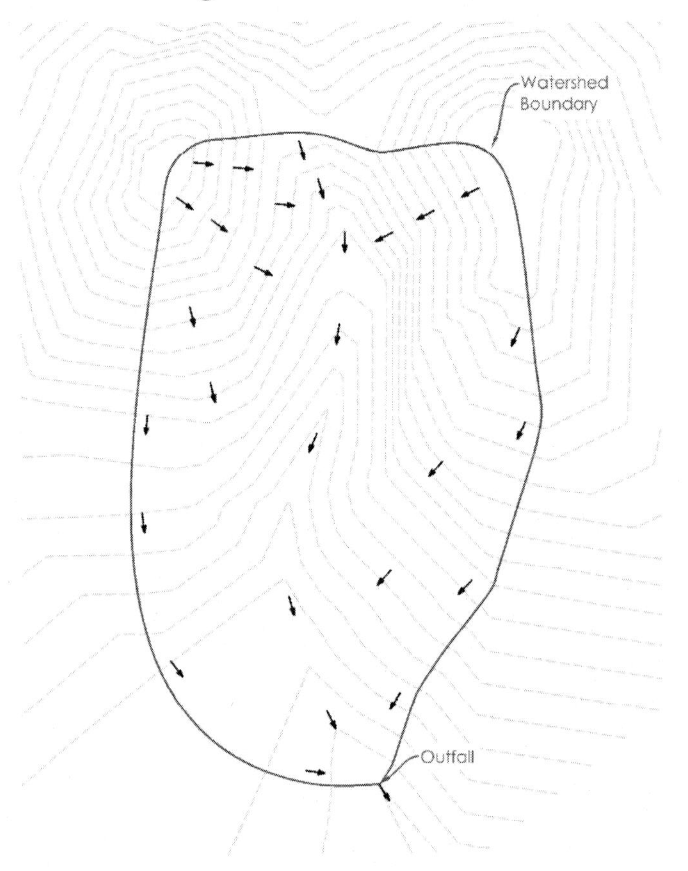

To understand the concept of hydraulic distance, imagine one planar sheet of rain drops fall on your watershed. Each drop will naturally travel to the outfall location. One drop that is the furthest distance from the outfall location sheet flows for 10 feet over asphalt and falls into a storm drainage inlet, where it is then carried through a series of pipes to the outfall location. Another drop, which starts much closer to the outfall, sheet flows through heavy underbrush and woods for 100 feet, concentrates and travels to the outfall. Though this rain drop started closer to the outfall, it will take longer to reach it, and is therefore at a more *hydraulically* distant location.

Time of concentration is important because it indicates the time at which runoff from every part of the watershed is contributing to the discharge at the outfall location. The longer the time of concentration, the lower the peak discharge rate at the outfall location and vice versa.

Hyetograph and Hydrograph

A hyetograph is a graph of the rainfall intensity over time for a storm event. A hydrograph is a graph of discharge over time. For stormwater design calculations, the hydrograph is typically used to represent the flow at the outfall locations. Note that the time to peak is directly correlated with the time of concentration (though these are not the same). Therefore, the peak of the hydrograph increases as the time of concentration decreases. Also note that the area under the hydrograph curve is the total volume of rainfall. Development increases the peak and total volume of runoff.

Storm Event

I won't go too in depth on how storm events are developed, but there are some general concepts that I want to touch on. When performing stormwater calculations, the regulations will specify the design storm events based on a return period. Typical design storm events include the 2-year storm event, 10-year storm event, 25-year storm event, 50-year storm event, and 100-year storm event. The year (i.e. 2-year) is the return period for that storm. In other words, the probability of that storm event occurring. For example, the 100-year storm event has a 1% chance of occurring in any given year. In addition to the return period, the storm event duration must also be considered. You will typically be required to analyze a 24-hour event.

So, for a 2-year, 24-hour storm event, you are analyzing a storm that has a 2-year recurrence of a specific rainfall accumulation in a 24-hour period. During storm events, rainfall intensities will vary throughout the 24-hour period. To accommodate this, you may analyze two types of storms: historic and synthetic. Historic storm events will be actual events that occurred in the region. Synthetic storm events are developed based on set parameters for your region.

Quality Control

When stormwater runs off of urban surfaces including parking lots, roads, and buildings, it brings pollutants with it. Some common stormwater pollutants include organic wastes, plant nutrients, and suspended solids. Organic wastes create a biological oxygen demand (BOD) that uses up dissolved oxygen needed for plant and fish life to survive. Plant nutrients include nitrogen and phosphorus that contribute to precipitous algae growth. Suspended solids often carry toxins and pathogens if they are not removed before entering the receiving water body. Most water quality BMP's are designed to remove suspended solids. I expect regulations to become stricter and more comprehensive as development continues to increase.

Calculations

Quantity Control

Time of Concentration

I previously explained the overall concept of time of concentration; here is an overview of the standard method for its calculation. This is the TR-55 method and is the most commonly used for stormwater calculations. The travel time is separated into categories; sheet flow, shallow concentrated flow, and channel flow.

In no case should the time of concentration be less than 6 minutes due to time-step increments used in the runoff calculations.

Sheet Flow

As stormwater first begins to run off the surface, it usually does so in the form of sheet flow. This means that the runoff is spread out over an area and moves along the surface like a sheet of water. Studies have shown that stormwater can maintain this method of travel for a maximum of 100 feet before it will start to concentrate into the next type of flow, shallow concentrated flow. Sheet flow typically accounts for most of the travel time in land development projects. Large-scale projects that analyze major watersheds may have more significant travel times in channel flow. The following equation is used to calculate sheet flow.

$$T_t = \frac{0.007(nL)^{0.8}}{(P_2)^{0.5}(S)^{0.4}}$$

- *T_t: Travel Time in Hours*
- *n: Manning's n Value (Represents the roughness of the surface)*
- *L: Travel Length in Feet (Not to exceed 100 feet)*
- *P_2: Rainfall for the 2-year, 24-hour storm event in Inches*
- *S: Slope of the surface*

Table 4-1: Manning's n Values for Sheet Flow[10]

Flow Rate	n
Smooth surfaces (concrete, asphalt, gravel, or bare soil)	0.011
Fallow (no residue)	0.05
Cultivated soils: Residue cover ≤20%	0.06
Cultivated soils: Residue cover >20%	0.17
Grass: Short grass prairie	0.15
Grass: Dense grasses	0.24

[10] (United States Department of Agriculture, 1986))

Flow Rate	n
Grass: Bermudagrass	0.41
Grass: Range (natural)	0.13
Woods: Light Underbrush	0.40
Woods: Dense Underbrush	0.80

Shallow Concentrated Flow

Shallow concentrated flow is flow over planar surfaces that begins to concentrate after about 100 feet of sheet flow. Flow is considered shallow concentrated flow until it reaches a swale, gutter, or pipe that carries it to the outfall.

$$T_t = \frac{L}{3600V}$$

- *L: Flow Length (ft)*
- *V: Average Velocity (ft/s)*

$$V = 16.13(S)^{0.5} \text{ (Unpaved)}$$

$$V = 20.33(S)^{0.5} \text{ (Paved)}$$

Channel Flow

Channel flow is calculated using Manning's equation:

$$V = \frac{1.49(R)^{\frac{2}{3}}(S)^{\frac{1}{2}}}{n}$$

This calculates the average velocity of the flow, so it can be used with the distance to calculate the travel time. This flow moves relatively quickly compared to sheet flow and is often a very small factor in typical land development projects.

Time of Concentration Calculation Example

Assume you have a flow path that extends for 2,350 feet. Fifty feet is sheet flow through woods with light underbrush at a slope of

2%. Three hundred feet is shallow concentrated flow over good grass at a slope of 4%, and 2,000 feet is pipe flow through an 18-inch concrete pipe system at a slope of 1%.

Sheet Flow Calculation

$$T_t = \frac{0.007(nL)^{0.8}}{(P_2)^{0.5}(S)^{0.4}}$$

- *T_t : Travel Time in Hours*
- *n: 0.40*
- *L: 50 ft*
- *P_2: 3.6 Inches*
- *S: 0.02 ft/ft*

$$T_t = \frac{0.007(0.40 * 50)^{0.8}}{(3.6)^{0.5}(0.02)^{0.4}} = 0.19\ hours = 11.4\ minutes$$

Shallow Concentrated Flow Calculation

$$V = 16.13(0.04)^{0.5} = 3.226\ ft/s$$

$$T_t = \frac{300}{3600(3.226)} = 0.026\ hours = 1.5\ minutes$$

Channel Flow Calculation

$$V = \frac{1.49(R)^{\frac{2}{3}}(S)^{\frac{1}{2}}}{n}$$

$$R = \frac{A}{P} = \frac{\pi(.75^2)}{\pi(1.5)} = 0.375$$

$$V = \frac{1.49(R.375)^{\frac{2}{3}}(.01)^{\frac{1}{2}}}{.011} = 7.75\ fps$$

$$t = \frac{d}{v} = \frac{2000\ ft}{7.75\ fps} = 4.73\ minutes$$

The total travel time will be the sum of these three components: 17.6 minutes.

Runoff (NRCS Curve Number Method)

When analyzing a proposed development for its stormwater impact, you will primarily be concerned with surface water runoff. There will be some cases, depending on location, where consideration must also be given to the recharging of groundwater aquifers, however, as this is rare and complex, we will focus solely on surface water runoff for the purpose of this text.

The runoff is the stormwater that does not infiltrate into soil and is not captured by plants or evaporated before it can contribute to an outfall. There are several methods for calculating stormwater runoff, but the NRCS Curve Number Method is the most common. The rational method is also used in small applications, but it is rare and a simple method, so I will let you defer to other texts for that information.

The NRCS Curve Number Method calculates the quantity of runoff using a variable called the *curve number*, which signifies the runoff potential of the surface in question.

Curve Number

The curve number is based on land use and is a measure of the runoff potential of a drainage area. The higher the curve number, the more the runoff potential. **Tables 4-2** provides a list of curve numbers from *Urban Hydrology for Small Watersheds*. As you can see, the impervious surfaces such as roads and buildings will have the highest curve number, and therefore the highest potential for producing runoff. Therefore, the transition of a site from a predominantly pervious condition to an impervious condition will have a significant impact on the quantity and speed of runoff. Once you've delineated your watershed, the next step in stormwater calculations is to identify the composite curve number for the drainage area.

One challenging aspect of runoff calculations is that two engineers will almost always come up with two different sets of numbers when analyzing a site. This can be frustrating when you are first learning, because there is no *right* answer and there is some subjectivity to the analysis. This is the case with several aspects of engineering, and a big difference from what you will have experienced in school. This is why experience and good judgment matter. This is also why it is so important to thoroughly understand the concepts. I will warn you that it is somewhat common in engineering to see experienced engineers work the numbers in their favor to avoid having an uncomfortable conversation with a client. I encourage you to resist this temptation. As a civil engineer, you have an ethical responsibility to produce an accurate analysis that is in keeping with the primary focus on health and safety.

Table 4-2: Curve Numbers[11]

Cover description		CN for HSG			
		A	B	C	D
Open space (lawns, parks, golf courses, cemeteries, etc.)	Poor condition (grass cover <50%)	68	79	86	89
	Fair condition (grass cover 50 to 75%)	49	69	79	84
	Good condition (grass cover >75%)	39	61	74	80
Impervious areas	Paved parking lots, roofs, driveways, etc. (excluding right of way)	98	98	98	98

[11] (United States Department of Agriculture, 1986)

Cover description		CN for HSG			
		A	B	C	D
Streets and roads	Paved; curbs and storm sewers (excluding right-of-way)	98	98	98	98
	Paved; open ditches (including right-of-way)	83	89	92	93
	Gravel (including right of way)	76	85	89	91
	Dirt (including right-of-way)	72	82	87	89
Western desert urban areas	Natural desert landscaping (pervious area only)	63	77	85	88
	Artificial desert landscaping (impervious weed barrier, desert shrub with 1- to 2-inch sand or gravel mulch and basin borders)	96	96	96	96
Urban districts	Commercial and business (85% imp.)	89	92	94	95
	Industrial (72% imp.)	81	88	91	93
Residential districts by average lot size	⅛ acre or less (town houses) (65% imp.)	77	85	90	92
	¼ acre (38% imp.)	61	75	83	87
	⅓ acre (30% imp.)	57	72	81	86
	½ acre (25% imp.)	54	70	80	85
	1 acre (20% imp.)	51	68	79	84
	2 acres (12% imp.)	46	65	77	82

Cover description		CN for HSG			
		A	B	C	D
Developing urban areas Newly graded areas (pervious areas only, no vegetation)		77	86	91	94
Woods-grass combination (orchard or tree farm)	Poor	57	73	82	86
	Fair	43	65	76	82
	Good	32	58	72	79
Woods [A]	Poor	45	66	77	83
	Fair	36	60	73	79
	Good	30	55	70	77
[A] Poor: Forest litter, small trees, and brush are destroyed by heavy grazing or regular burning; Fair: Woods are grazed but not burned, and some forest litter covers the soil; Good: Woods are protected from grazing, and litter and brush adequately cover the soil.					

Hydrologic Soil Group (HSG)

As you can see in **Table 4-2**, the HSG must be known to identify the correct curve number. If the surface is not impervious, the soil characteristics will control the quantity of runoff. Soils are broken down into four hydrologic categories which are used to identify the runoff potential; *A*, *B*, *C*, and *D*. **Table 4-3** provides an overview of each category.

Table 4-3: Hydrologic Soil Groups

HSG	Description
A	• Low runoff potential, water transmitted freely • 10% clay, 90% sand/gravel
B	• B Type Soils • Moderately low runoff potential • 10-20% clay, 50-90% sand
C	• Moderately high runoff potential • 20-40% clay, <50% sand
D	• High runoff potential • >40% clay, <50% sand
A/D, B/D, C/D	• In some cases, you will see a soil type such as A/D. In this case, the soil type allows for good drainage, but the water table inhibits drainage. For this reason, it is safer to assume these types of soil fall in the D category.

NRCS Runoff Equations

The NRCS runoff equations can be used for areas of up to 2,000 acres, which will cover most of the projects you work on unless you are analyzing very large-scale watersheds. Note in the following equations that the curve number is used to calculate the maximum soil retention, which is in turn used to calculate the initial abstraction. Therefore, the only parameters needed to calculate the runoff are the curve number and rainfall amount.

$$Q = \frac{(P - I_a)^2}{(P - I_a) + S}$$

- *Q = Runoff (inches)*
- *P = Rainfall Amount (Inches)*
- *I_a = Initial Abstraction (storage, interception, infiltration, evaporation)*
- *S = Maximum Soil Retention*

$$I_a = 0.2S$$

$$Q = \frac{(P - 0.2S)^2}{(P + 0.8S)}$$

$$S = \frac{1000}{CN} - 10$$

$$CN = \frac{1000}{(10 + 5P + 10Q - 10(Q^2 + 1.25QP)^{.5})}$$

Runoff Calculation Example Problem

Let's say we have identified a watershed that is 100 acres. We have determined that 20 acres is made up of roads, parking areas, and buildings; 25 acres is made up of woods on type C soil; 25 acres is made up of woods on type A soils, and the remaining 30 acres is made up of good grass on type A soils. Calculate the volume of runoff from 3.6 inches of rainfall.

First, we need to select the curve number for each area:

Land Use	Acres	Soil Type	Curve Number
Roads/Parking Lots	20	N/A	98
Woods (Fair)	25	C	73
Woods (Fair)	25	A	36
Good Grass	30	A	39

Now we will calculate the composite curve number for the entire drainage area:

$$\frac{\sum(Acres * Curve\ Number)}{\sum Acres} = \frac{5855}{100} \approx 59$$

Using the curve number, we can calculate the total runoff:

$$S = \frac{1000}{CN} - 10 = \frac{1000}{59} - 10 = 6.949$$

$$I_a = 0.2S = 0.2(6.949) = 1.390$$

$$Q = \frac{(P - I_a)^2}{(P - I_a) + S} = \frac{(3.6 - 1.390)^2}{(3.6 - 1.390) + 6.949} = .53\ inches\ of\ runoff$$

This calculation provides us with the total volume of runoff. In most cases, you will also need to determine peak flow rates at the outfall for various storm events. The calculations associated with this are beyond the purview of this text and are almost always completed using modeling software. I do encourage you to investigate manually calculating resultant hydrographs if you want to have a firmer understanding of this concept. I've read several books on the topic, but I can tell you there is no better way to teach yourself the calculation methodology than doing it yourself.

Design Elements

Storm Drainage Infrastructure

Storm drainage infrastructure, such as pipes, catch basins, and swales will route the stormwater to a stormwater management facility or outfall location. Whether you're grading a road, parking lot, or grassed area, you will want to make sure you identify where the stormwater will be collected and how the pipe network will progress toward the outfall. This will determine catch basin locations, pipe sizes, and pipe depths. If you don't consider the storm drainage infrastructure during your grading design, you may wind up with very deep pipes or situations where you will have no way of getting some of the stormwater where you want it.

Pipe Material

Most of the stormwater pipes you specify in stormwater distribution systems will be reinforced concrete pipe (RCP). You may use also materials such a corrugated plastic pipe (CPP) or corrugated metal pipe (CMP), but this is much less common.

Pipe Sizing and Minimum Slope

Basic pipe sizing is done using Manning's equation. The level of detail used for storm system design will vary depending on the size of the project and requirements of the municipality. In some cases, basic spreadsheet calculations will be used, but in most cases, a full software model should be developed. The minimum pipe size is for design is usually 15-inches, though smaller pipes can be used in some cases. Pipe sizes on a typical project will usually range from 15-inches up to 60-inches. Large stormwater projects further downstream in the system may require much larger stormwater conveyance infrastructure. Following are examples of minimum slope and capacity calculations for storm drainage pipes.

Minimum Slope

The flow must maintain a minimum velocity of 2 feet per second to prevent deposits from settling on the bottom. This is called the self-cleaning velocity, and it is calculated using Manning's equation with a typical roughness factor of 0.013. **Table 4-4** shows the minimum slopes recommended by the *Ten State Standards*. Following is a sample calculation for a 10-inch pipe. Note that the recommended values provided by the *Ten State Standards* are a little more conservative since this calculation assumes full pipe flow:

$$V = \frac{1.49}{n} R^{\frac{2}{3}} S^{\frac{1}{2}} \rightarrow S = \left(\frac{V}{\frac{1.49 R^{\frac{2}{3}}}{n}} \right)^2$$

$$R = \frac{A}{P} = \frac{\pi r^2}{\pi d} = \frac{\pi * \frac{5}{12}^2}{\pi * \frac{10}{12}} = 0.351$$

$$S = \left(\frac{V}{\frac{1.49 R^{\frac{2}{3}}}{n}} \right)^2 = \left(\frac{2}{\frac{1.49(0.351)^{\frac{2}{3}}}{0.013}} \right)^2 = \mathbf{0.25\%}$$

Pipe Capacity

The pipe capacity can also be calculated using Manning's equation. Following is a sample calculation of the capacity for a 10-inch pipe at a 1% slope. Note that the slope can have a significant impact on the capacity.

$$V = \frac{1.49}{n} R^{\frac{2}{3}} S^{\frac{1}{2}} = \frac{1.49}{0.013} * 0.351 * \left(0.01^{\frac{1}{2}}\right) = 4.03\ ft/s$$

$$Q = VA = 4.03 * 0.55 = 2.20\ cfs = \mathbf{986.03\ gpm}$$

Table 4-4: Minimum Slope[12]

Diameter (inches)	Minimum Slope (%)
8	.40
10	.28
12	.22
15	.15
16	.14
18	.12
21	.10
24	.08
27	.067
30	.058
33	.052
36	.046
39	.041
42	.037

[12] (Greater Lakes - Upper Mississipi RIver Board of State and Provincial Public Health and Environmental Managers, 2014)

These are absolute minimum slopes and assume a pipe full flow condition. Often your firm will have a general standard minimum slope such as 0.005 ft/ft. The reason for this is that though lower slopes are possible in special conditions, caution should be taken to ensure that the slope is constructible within reasonable tolerances. It should also be noted that the steeper the slope, the higher the capacity, so using the minimum slope should be avoided if possible.

Minimum Cover

The standard minimum cover requirement for a storm drainage pipe (RCP) is 1 foot. This may vary depending on the loading conditions above the pipe. During the design process, it is important to give yourself a little bit more than one foot of cover to allow for variations in the grade over the pipe and construction tolerance. If you design a pipe with exactly one foot of cover, it is possible that the installed pipe will have less than one foot of cover in some areas and you will not receive the necessary close-out from the municipality.

Minimum Separation

The standard minimum vertical separation between the storm drainage pipe and other crossing pipes is 18 inches. Caution should also be taken when you have large storm drainage pipes above other pipes due to the additional vertical load that this will place on the crossing pipe.

Easements

Easements will be required in any case where the public entity will maintain ownership and maintenance responsibilities of the stormwater infrastructure. A typical easement width is fifteen feet, but this will vary depending on agency guidelines, depth, and size. The main will need to be accessible in any case where maintenance is needed, and an adequate easement is needed to allow for this maintenance.

Anti-Seep Collar

Anti-seep collars are used whenever there is a potential for the seepage of water to undermine the interface between the pipe and the soil. The collars are designed to increase the distance that the water must travel along the pipe. There is not a lot of consensus regarding the efficacy of this method, but it is still commonly used in design. In stormwater applications, these are often used for the outlet pipe of a stormwater basin as shown in **Figure 4-3**. They are also used in areas where there is a high groundwater table or saturated soils such as wetlands or stream crossings.

Culvert

Culverts are typically used to convey water under roads and driveways. In smaller applications, pipes are used. In larger applications, box culverts are often used where possible. If the area needed to maintain proper flow under the road is too large for these applications, a bridge is needed. I won't go into detail regarding culvert design in this text. However, I will point out that it is more complex than the simple application of Manning's equation discussed previously. Culverts will have varying inlet and outlet controls due to capacity and tailwater conditions that increase the complexity of the culvert sizing calculations. The Federal Highway Administration (FHWA) has the most comprehensive guidelines for culvert sizing.

Structures

In a storm drainage system, structures will be needed in any areas where flow is expected to enter the system and where there is any bend in the system alignment. Following are the various types of structures that are used. These are depicted in **Figure 4-2**.

Curb Inlet

A curb inlet is placed along curbs on roadways and in parking lots to catch stormwater flowing along the curb line or directed to a low point in the parking area.

Combination Inlet

A combination inlet consists of a curb inlet with a grate inlet. These are often used in parking lots. You will see them used in roadway applications, but great care must be taken to ensure that the grate is not within the vehicle path. Over time the grates will settle and can create a serious road hazard.

Grate Inlet

A grate inlet is placed at a low point in a grassed or paved area to collect stormwater. If these are used in grassed areas where leaves and debris are prevalent, screening should be used to facilitate cleaning or the grate.

Yard Inlet

A yard inlet is placed in grassed areas at low points or at the discharge point of swales.

Junction Box

A junction box is used in areas where it is necessary to change direction, but not collect stormwater.

Swale

A swale is typically a grassed trapezoidal conduit used to transport stormwater along the surface. Swales can also be paved if steep slopes make stability of a grassed swale infeasible.

Subsurface Drain

In cases where groundwater is a concern, it may be necessary to provide a permanent drain to ensure the groundwater does not negatively impact the foundation, surface parking, or driving area. It is preferable to defer to a geotechnical engineer for advice on subsurface drainage.

Level Spreader

A level spreader takes concentrated flow from a pipe or swale and converts it to sheet flow to prevent erosion from high local velocities.

Stormwater Management Infrastructure

Detention Basin

The detention basin (dry basin) is the most commonly used stormwater management structure. This is largely due to its relatively low cost, ease of construction and maintenance, and ability to treat large amounts of stormwater runoff. A detention basin is designed to attenuate peak flow rates by detaining excess stormwater and releasing it over an extended period. Detention basins are often used as sediment basins during construction, then converted to permanent water quality basins after the site has been stabilized.

Figure 4-2: Storm Drainage Inlets

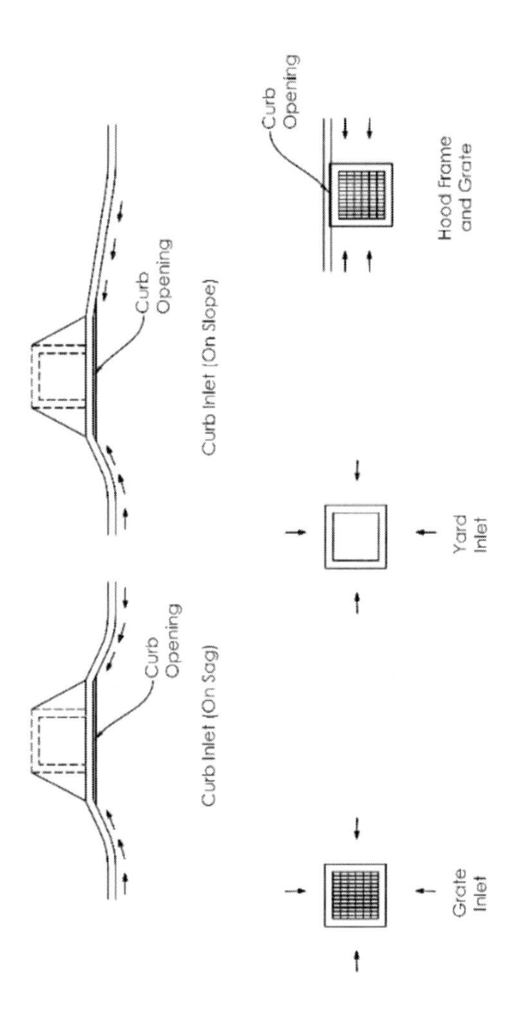

The performance of a detention basin depends on the inflow rate of stormwater, its shape and size, the underlying soils, and the outlet structure and emergency spillway configurations. Detention basin design is an iterative process, but a solid understanding of the fundamental concepts will be a big help in coming up with an efficient and effective design. **Figures 4-5** and **4-6** show a typical detention basin plan and cross-section.

Underlying Soils

The infiltration rate of the underlying soils in a detention basin can have a large impact on its performance. However, most jurisdictions have restrictions regarding use of this parameter in water quantity analysis calculations.

Shape and Size

The shape and size of the basin will impact its ability to facilitate water quality treatment and attenuate peak flows. The surface area influences the settling of particles while the stormwater is detained in the basin (the *In-Depth: Understanding Orifice Discharge* section explains this in more detail). The shape of the basin will determine the travel time of the water from when it enters the basin to when it exits through the outlet structure. You want this travel time to be as long as possible to allow for the settling of particles as the water flows through the basin. For this reason, it is generally required that the length be twice as long as the width. If this is not possible, or if the inlet and outlet are not able to be located to maximize the distance of travel, you can use flow diversion baffles to facilitate an increase in travel time.

Though there are general rules of thumb and very simple equations that can be used to preliminarily size a basin, I highly recommend creating a general computer model using as much data as is available. There are too many variables for most rules of thumb to provide a reasonably accurate basin size.

Outlet Structure

Orifice and weir locations and sizes will have an impact on how quickly stormwater is released from the detention basin. The typical outlet structure will have a water quality orifice located at the bottom. This allows the water quality storm (usually a storm that produces one inch of runoff) to discharge over of a relatively long period of time, between 24 and 72 hours, to allow for the settling of particles that may carry pollutants. Mid-range orifices are often added to mitigate the post-development peak flow rates to pre-development conditions. It is optimal to have the entirety of the water quality storm discharge through the low-flow water quality orifice at the bottom of the outlet structure. Mid-range orifices can be used to prevent the 10-year storm from over-topping the outlet structure. The 25-year storm can discharge through the top of the outlet structure, modeled as a rectangular weir. A typical outlet structure is shown in **Figure 4-3** and additional detail regarding orifice structure design is included in the *In Depth: Understanding Orifice Discharge* section.

Figure 4-3: Outlet Structure (Detention Basin)

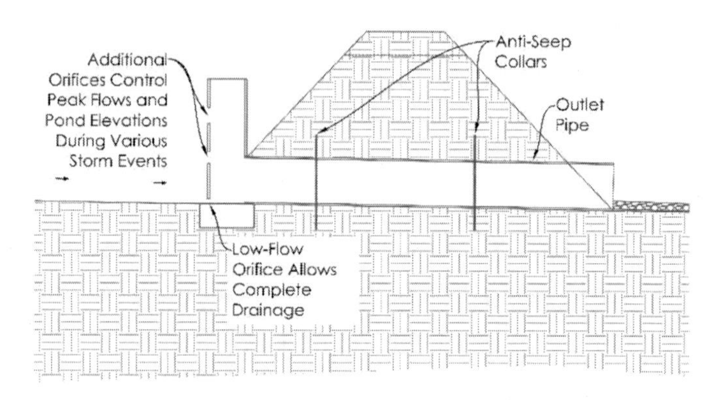

Emergency Spillway: An emergency spillway, which is a rip-rap protected outlet in the pond dam (typically one to one and a half feet below the top of the pond dam), allows the 100-year storm to safely pass without undermining the dam. This should be located in a cut section if possible. The exclusion of the emergency spillway will result in erosion and possible failure of the pond dam during high storm events.

In Depth: Understanding Orifice Discharge

One key factor when designing a detention basin is meeting the necessary requirements for dewatering the water quality volume. The water quality volume will vary depending on the jurisdiction, however, it is typically the first inch of runoff for a dry basin and the first half inch of runoff for a wet basin. This is often called the first flush, and this volume must take a minimum of 24 hours to discharge from the basin to meet typical water quality requirements. When you are designing a basin, you will often run into situations where you are having trouble meeting this requirement. It is helfpul to understand some basic concepts related to orifice discharge in order to determine the most optimal basin design. First let's look at the basic orifice equation:

$$Q = CA\sqrt{2gh}$$

- *Q: Flow Rate (cfs)*
- *A: Orifice Diameter (ft)*
- *g: Gravity (32.2 ft/s2)*
- *h: Height (ft)*
- *C: Orifice Coefficient (0.6 for circular orifice)*

Q is the flow rate in cubic feet per second. In other words, Q represents the change in volume over time. For simplicity, we will assume the surface area of the basin will remain the same as the height varies. We know this is not true since the surface area in the basin will actually vary with the height, but the general concept can be more easily understood if we use this assumption.

So we can rewrite Q:

$$Q = \frac{\Delta V}{\Delta t}$$

And since the Area is constant, only the height will vary with time. Therefore, we can rewrite this equation:

$$Q = \frac{\Delta V}{\Delta t} = -A\frac{\Delta h}{\Delta t}$$

The negative sign must be added since the change in height will be negative and our flow rate must be positive. Now let's plug this back into our orifice equation:

$$-A\frac{\Delta h}{\Delta t} = Ca\sqrt{2gh} \rightarrow \frac{\Delta h}{\Delta t} = -\frac{Ca\sqrt{2gh}}{A}$$

Now we are set up to create a differential equation. We want to break the changes into small increments, so we can rewrite this equation as follows:

$$\frac{dh}{dt} = -\frac{Ca\sqrt{2gh}}{A}$$

This is a separable differential equation, so we can isolate the variables:

$$\frac{dh}{\sqrt{h}} = -\frac{Ca\sqrt{2g}}{A}dt$$

Let's simplify this by saying that $k = -\frac{Ca\sqrt{2g}}{A}$ since we know this will be constant. Now we are ready to integrate and solve for time.

$$\int_{h_i}^{h_f} h^{\frac{1}{2}}dh = -k\int_{t_i}^{t_f} dt$$

$$2\sqrt{h} = -kt + C_1$$

Now we need to solve for C_1 at the initial condition of $t = 0$:

$$C_1 = 2\sqrt{h_i} + k(0) \rightarrow C_1 = 2\sqrt{h_i}$$

Therefore:

$$2\sqrt{h_f} = -kt + 2\sqrt{h_i}$$

Solving for t:

$$t = \frac{2\sqrt{h_f} - 2\sqrt{h_i}}{-k}$$

Plugging our constants back in for *k* we get:

$$t = \frac{-2A(\sqrt{h_f} - \sqrt{h_i})}{Ca\sqrt{2g}}$$

This equation will give you the time it takes to dewater the basin to any given final height. It is not necessary to remember how to derive the equation. The derivation is just meant to provide an understanding of the progression from the basic orifice discharge equation to an equation for dewatering time. This may be helpful to some people, but also may cause confusion for others. If you are in the latter group, the important thing is that you focus on understanding the next part rather than dwelling on the mathematics.

Now let's look at the impact that orifice size, height (depth), and area have on the dewatering time. Note that these are the only three variables, and the height can only be modified if the area is modified; as the area increases, the height will decrease since the water quality volume is constant. So, let's look at some scenarios to determine the impact that changes to the area and orifice size will have on the dewatering time. I'll run through these rather quickly, but I encourage you to run the calculations yourself to better understand the concept.

Let's assume we have a water quality volume of 30,000 cubic feet. We will start with a basin that has a surface area of 10,000 square feet, so the height of the water quality volume in the basin will be 3 feet. If we start with a diameter of 6 inches for our water quality orifice, the dewatering time will be approximately 14 hours. If we decrease the orifice diameter to 3 inches, we see that the dewatering time increases to approximately 58 hours. Note that even though this adjustment would allow you to meet the requirements for dewatering, this assumes that all of the water flows through the water quality orifice. If the height of the water in the basin is 3 feet, you will need to place any additional orifices above this elevation. This may be an issue if you have a relatively shallow basin. A graph showing how the dewatering time varies with the orifice diameter can be found in **Figure 4-4**.

Now let's look at the same conditions, but we will adjust the surface area instead. In this case we will assume that we reduced the orifice diameter to the minimum 4 inches, but an initial surface area of 5,000 square feet is giving us a height of 6 feet and dewatering time of only 23 hours. If we increase the surface area to 15,000 square feet, we can see that the height reduces to 2 feet and the dewatering time increases to 40 hours. Note that even though the height decreased, there is still a large increase in dewatering time. If you look back at the equation we derived, you can see that the height is both a much small number and taken to the ½ power, so it has less impact on the overall time. A graph showing how the dewatering time varies with surface area can be found in **Figure 4-4**.

Pond design will always require some iteration. However, if you understand these basic concepts, it will make the iterative process much easier and help you design a proper basin that meets the requirements in an efficient cost-effective manner.

Retention Basin

A retention basin (or wet pond) is similar to a detention basin except that it is designed to hold a permanent pool of water. This permanent pool is helpful in treating water quality. In performing a hydrologic analysis, it is assumed that the bottom of the pond is equal to the water surface elevation of the permanent pool. This is controlled by the location of the orifice in the outlet structure. Underlying soils

Figure 4-4: Dewatering Time

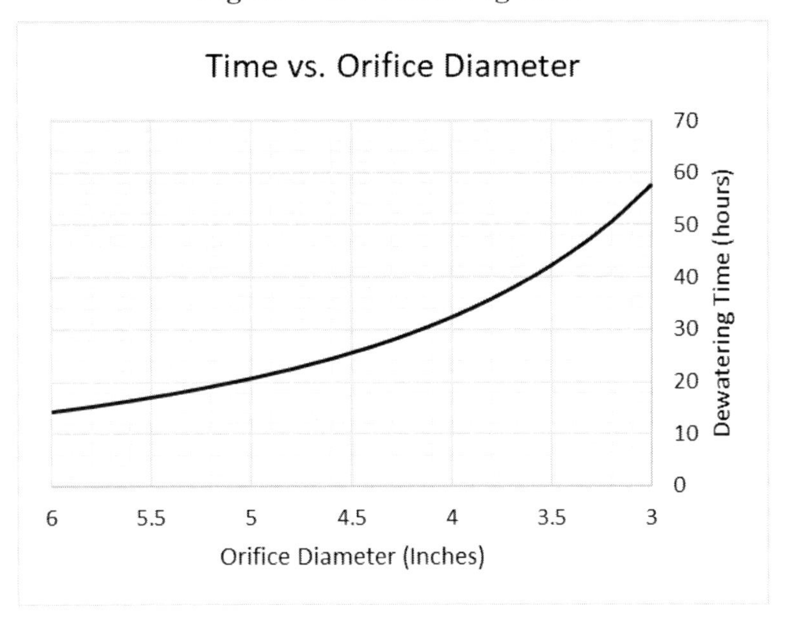

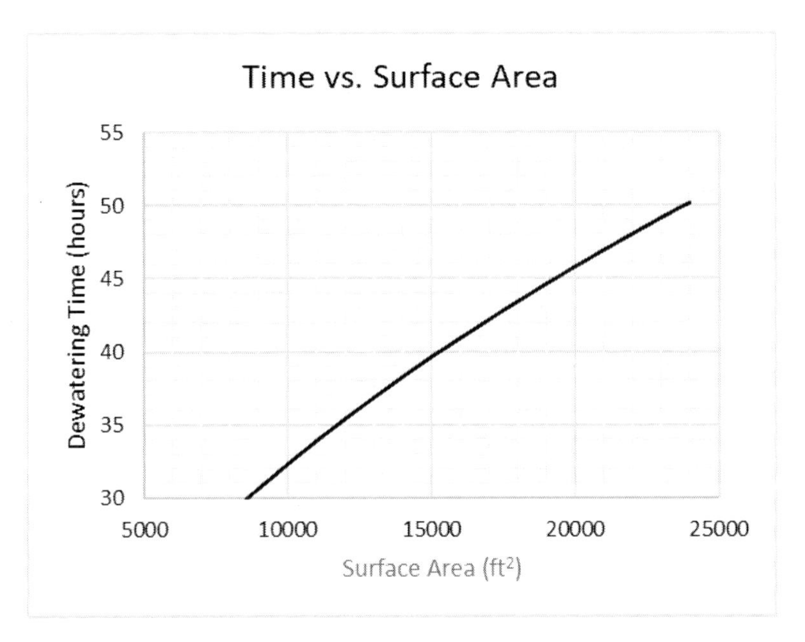

should be investigated to ensure that the infiltration rate is not too high to prevent the permanent pool from remaining. If the in-situ soils are not adequate, the use of a geomembrane or clay layer at the bottom of the pond may be necessary to maintain the permanent pool. The permanent pool should maintain a minimum depth of four feet to allow for proper vegetative growth and prevent conditions that promote mosquito infestations. Reference can be made to **Figure 4-6** for a depiction of a typical Retention Basin cross-section.

Infiltration Basin

An infiltration basin is sometimes referred to as a No Discharge Basin, because it is designed to infiltrate any runoff from most storm events. Different jurisdictions will have varying requirements on whether the infiltration basin must hold the 10-year, 25-year, or 100-year storm. These basins will typically only have an emergency spillway. It is very important that you have a solid understanding of the soil and groundwater conditions in the area where the basin will be constructed. Geotechnical testing to determine infiltration rates should always be completed when designing an infiltration basin. Otherwise, you will likely have significant issues in the field.

Bioretention Cell

Bioretention systems are intended to replicate existing hydrologic conditions after site development. As was discussed previously, urban development results in the addition of impervious surfaces that modify hydrologic conditions and increase pollutant loads. Bioretention systems can be used to capture the runoff from these impervious surfaces. The typical bioretention cell consists of

Figure 4-5: Typical Detention Basin

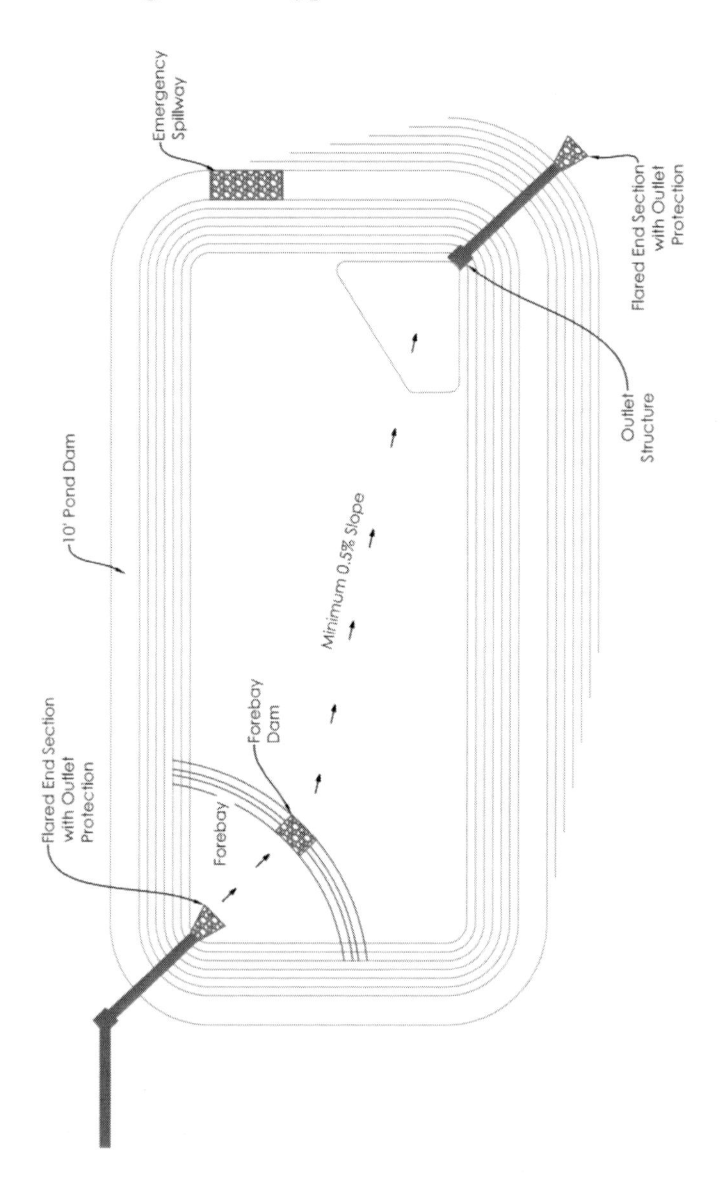

Figure 4-6: Basin Cross Sections

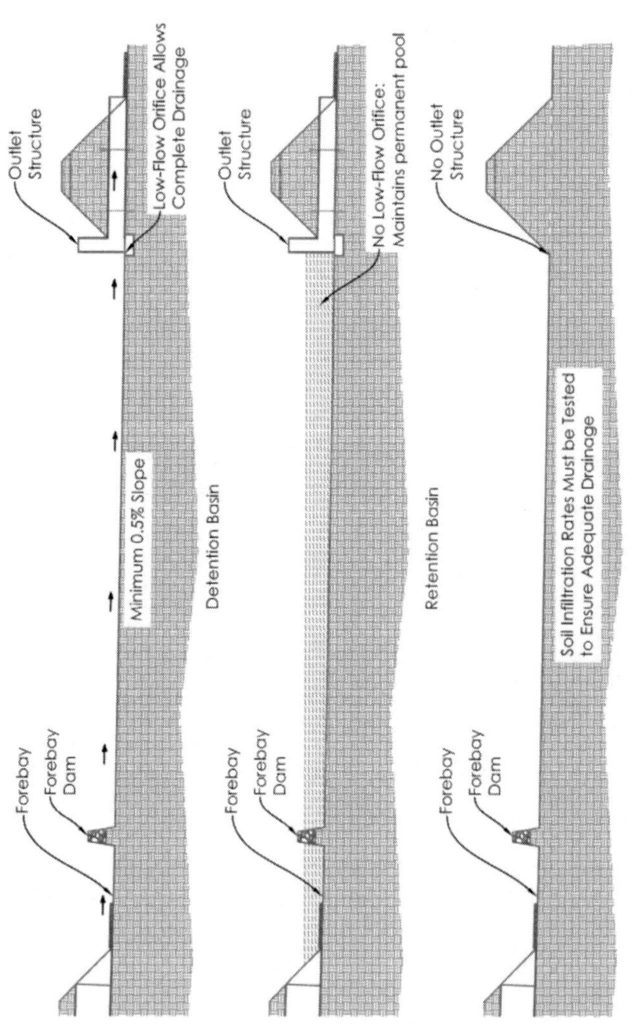

specified types of plants, a filter media layer, and an underdrain system depending on the infiltration rate of the underlying soil. Some states always require the use of an underdrain due to typically low infiltration rates of in-situ soils.

Bioretention is mostly used in urban environments where space and aesthetics are a concern. It is not effective in large scale applications and is usually used to capture runoff from a maximum of two acres. Therefore, it is commonly used to capture runoff from parking lots, buildings, or sidewalks. The excess stormwater runs off of these impervious surfaces and flows in to the ponding area of the bioretention cell. This ponding area is usually six to nine inches in depth. The stormwater then infiltrates through the filter media, which is an engineered mix that allows for rapid infiltration. The efficacy of the filtration for water quality treatment varies. Stormwater characteristics and filter media should be carefully analyzed during the design process. Once the stormwater filters through the soil it either infiltrates through the underlying soil or is discharged through the perforated underdrain. **Figure 4-7** shows a basic Bioretention Cell configuration.

Figure 4-7: Bioretention Cell

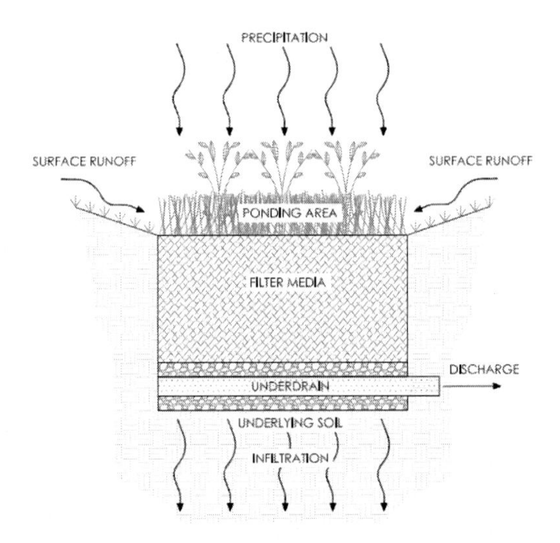

Design Process

The stormwater conveyance design process varies greatly from project to project. The first step is to identify and thoroughly review your design requirements, including the following:

- *Allowable Calculation Methodologies*
- *Quantity Control Requirements*
- *Quality Control Requirements*

These requirements have been discussed in some detail in the previous sections. Just remember that these will likely be different depending on the location of your project, so it is extremely important that you review the local stormwater design standards before beginning.

Stormwater design involves the site grading, stormwater conveyance infrastructure, and stormwater management infrastructure. All of these components must be considered from the start to determine the most optimal design. Since reference can be made to previous sections for additional detail on the following steps, I will list the tasks in very general terms.

- ***Step 1: Identify regulatory requirements***
- ***Step 2: Calculate the pre-development peak flow rate(s)***
    - *Delineate pre-development watershed(s)*
    - *Determine pre-development time of concentration and curve number(s)*
    - *Calculate the pre-development peak flow rate(s)*
- ***Step 3: Complete preliminary site grading and storm drainage infrastructure layout***
- ***Step 4: Preliminarily size stormwater management structures***
    - *Delineate post-development watershed(s)*
    - *Determine post-development time of concentration and curve number(s)*
    - *Preliminarily design stormwater management structures*
    - *Calculate post-development peak flow rate(s)*
    - *Adjust stormwater management structures as necessary to meet pre/post peak flow requirements*
- ***Step 5: Refine grading and storm drainage infrastructure layout***
    - *Complete final grading*
    - *Finalize all storm drainage infrastructure location and sizing*
- ***Step 6: Recalculate pre/post development flow rates***
    - *Revised storm drainage calculations as necessary*
    - *Verify that requirements are met*
    - *Make adjustments to stormwater management structures if necessary to meet requirements*

Figure 4-8: Storm Drainage Plan (Commercial)

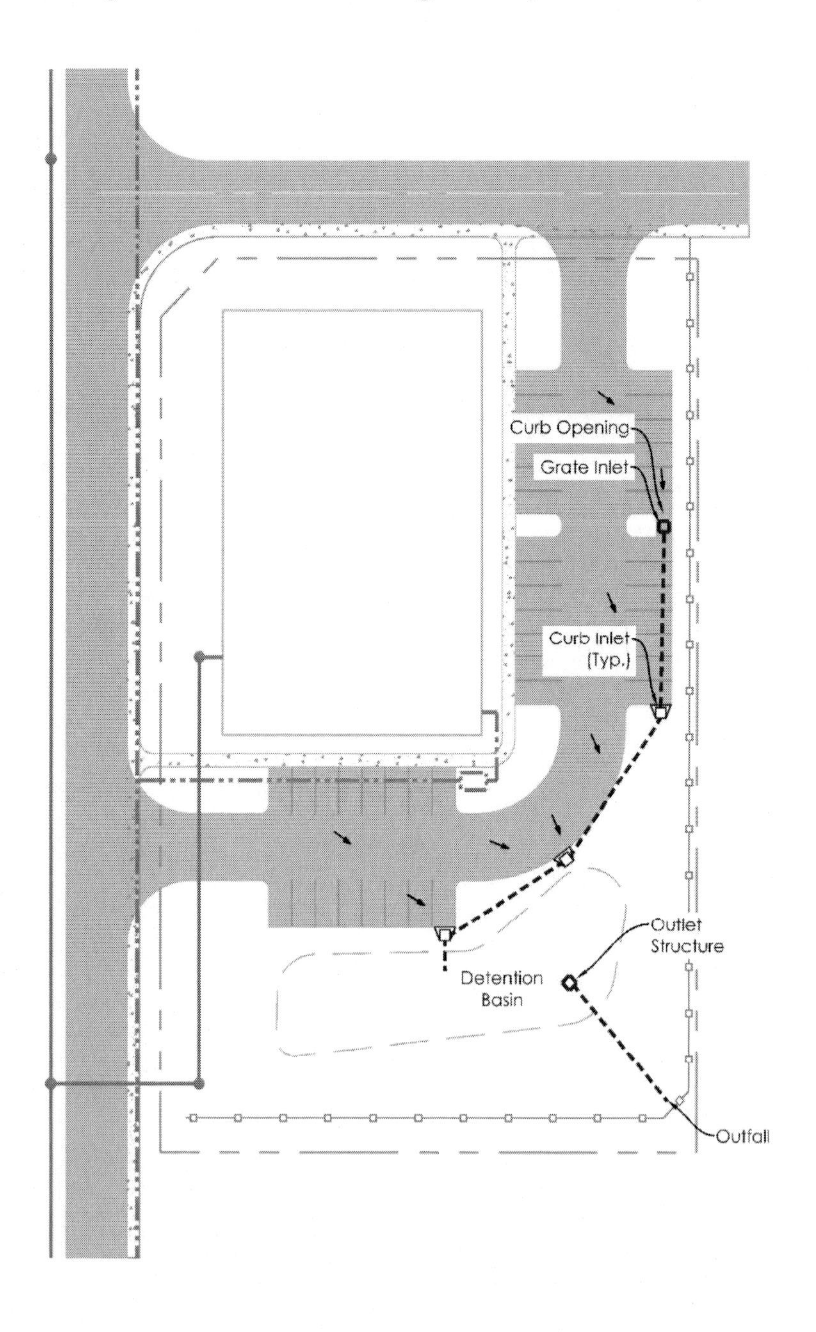

Figure 4-9: Storm Drainage Plan (Residential)

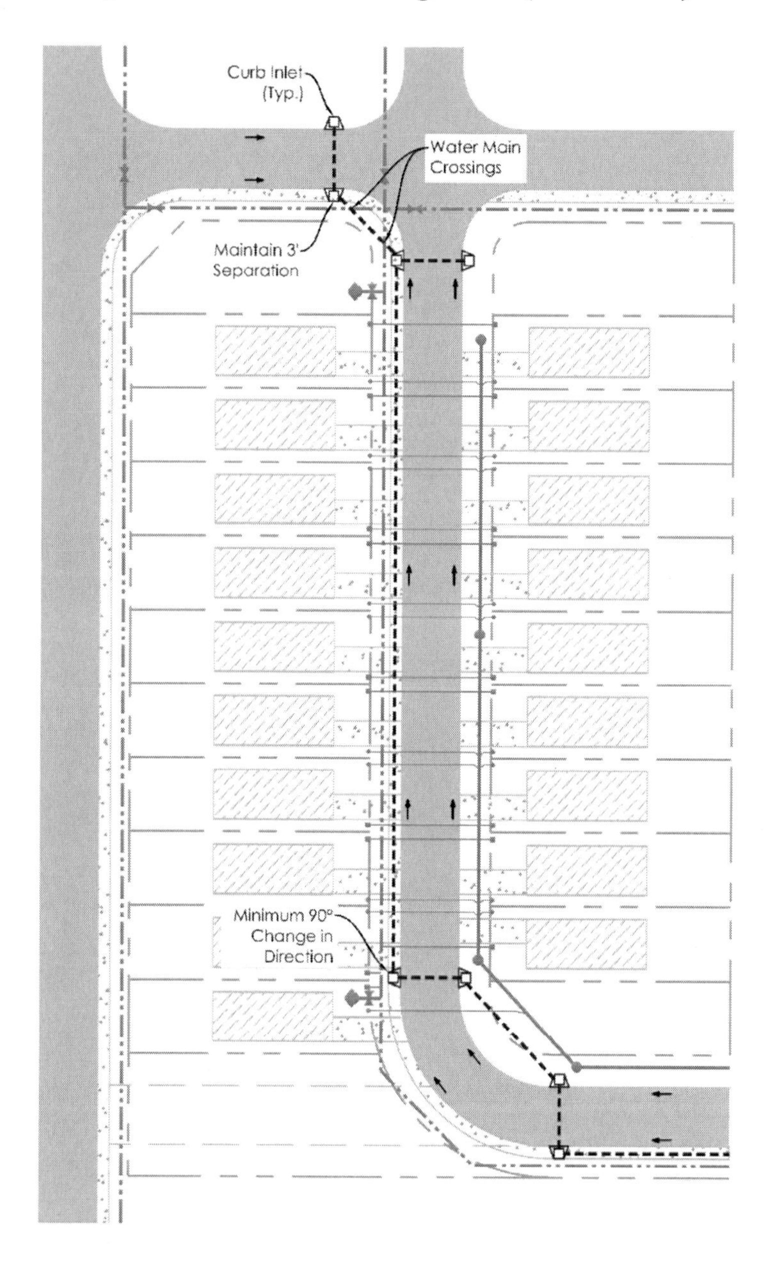

Figure 4-10: Storm Drainage Profile

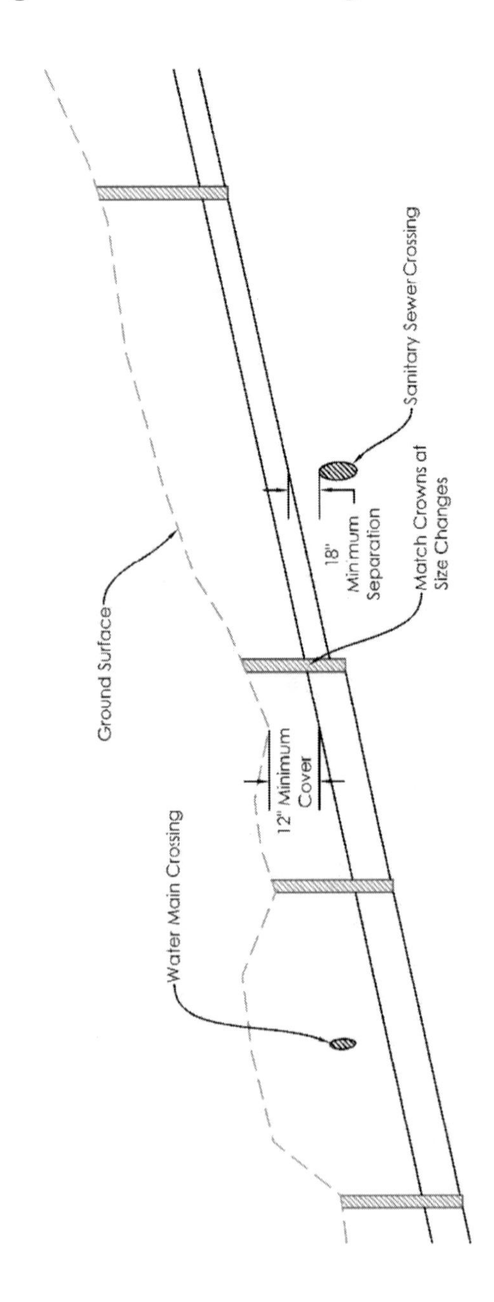

Five
Erosion & Sediment Control

The Big Picture

During mass grading operations, erosion and sediment control should be carefully designed. The site will largely be stripped of the trees and vegetation that prevent erosion and sediment transport under normal conditions. The goal of erosion and sediment control design is to keep this sediment from leaving the construction site. This is usually accomplished using several erosion and sediment control best management practices (BMP's).

Remember to think big when you are designing for erosion and sediment control. Your design includes the selection of best management practices to prevent sediment from leaving your site and to prevent erosion of any areas potentially impacted by the new construction. The site conditions will vary during project grading, so multiple erosion and sediment control plans may be necessary as construction progresses.

Calculations

There are numerous calculation methods for erosion and sediment control. This is because each BMP has its own guideline for calculation of its efficiency in removing sediment. A removal efficiency of 80% is standard, but some local regulations may be stricter. I won't go into detail regarding calculation methods here, but you can refer to the Department of Health and Environmental Control for calculation methods related to each BMP.

Design Elements

Sediment Basin

Any project that consists of greater than 10 acres of disturbed area discharging to a single point will require the use of a sediment basin. A sediment basin is designed to remove as much sediment as possible from stormwater runoff before it leaves the project site. **Figures 5-1** and **5-2** illustrate the components of a typical sediment basin. Following is a discussion of components and considerations that make up an effective sediment basin design.

Surface Area

Surface area is the most important design consideration for an effective sediment basin. Settling is based on gravity, so particles will settle slowly, and it is important that a large number of particles settle prior to reaching the outlet. This means that the longer each particle takes to reach the outlet, the further it will settle. For this reason, the length to width ration should be a minimum of 2:1. Flow direction baffles can be used in cases where the flow path from the inlet to the outlet is not adequate.

Skimmer and Outlet Structure

DHEC now requires a skimmer to be used on all sediment basins. A skimmer, as the name implies, skims water from the surface and carries it to the orifice located at the bottom of the outlet structure. This is effective at reducing sediment discharge because the surface water will have less sediment as it will have had some time to settle. In most cases, the sediment basin will be converted to a detention basin after construction. Therefore, the outlet structure will be converted to the detention basin configuration once the site has been stabilized. Prior to this, all orifices, other than the low-flow orifice that the skimmer is tied into, should be closed.

Sloped Pond Bottom

If the basin bottom is graded flat, stormwater will not fully discharge to the outfall. For this reason, the bottom of the basin must be sloped at a minimum of 0.5% to allow for positive drainage to the outlet structure.

Forebay

The forebay provides an area for sediment to settle prior to entering the main basin. This leaves a smaller area for the contractor to clear of sediment during construction.

Cleanout Stake

The cleanout stake provides the contractor with an indicator of when the sediment basin needs to be cleared of sediment.

Porous Baffles

Porous baffles are installed perpendicular to the flow path in a sediment basin to slow and spread the flow across the basin. Contrary to popular belief, these are not intended for filtration. The slowing and spreading of the flow facilitates particle settling.

Figure 5-1: Sediment Basin

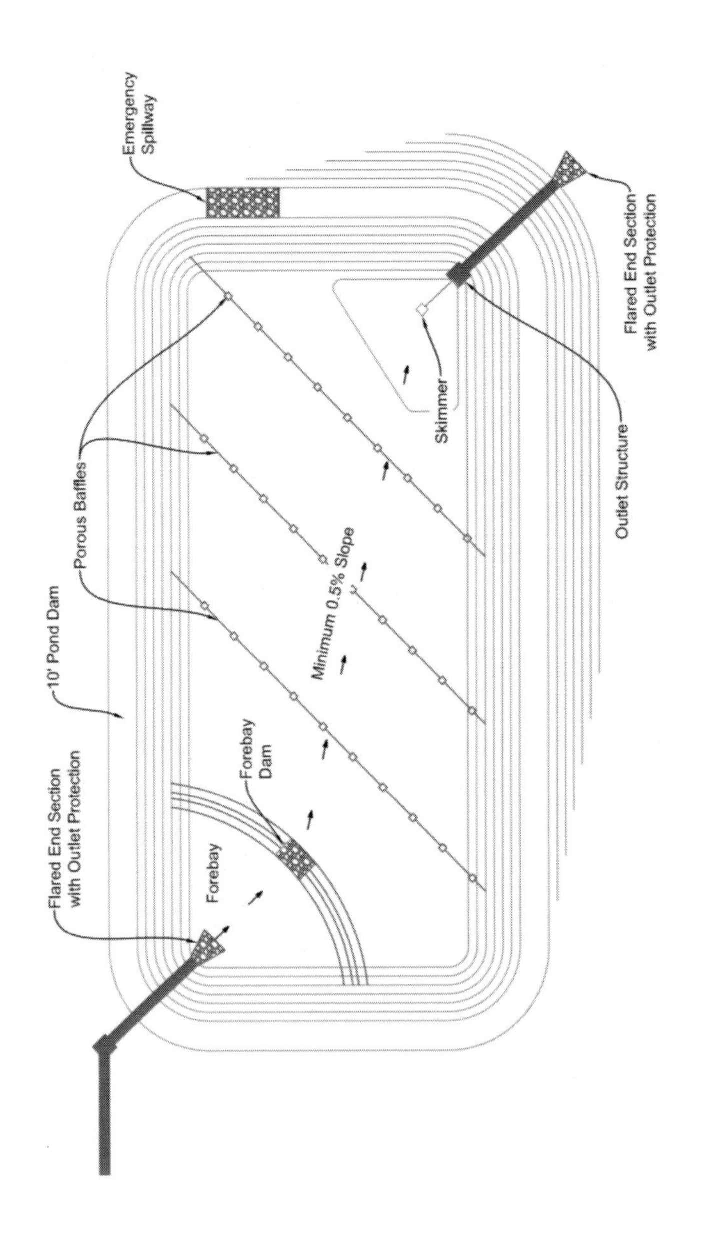

Figure 5-2: Sediment Basin Cross Section

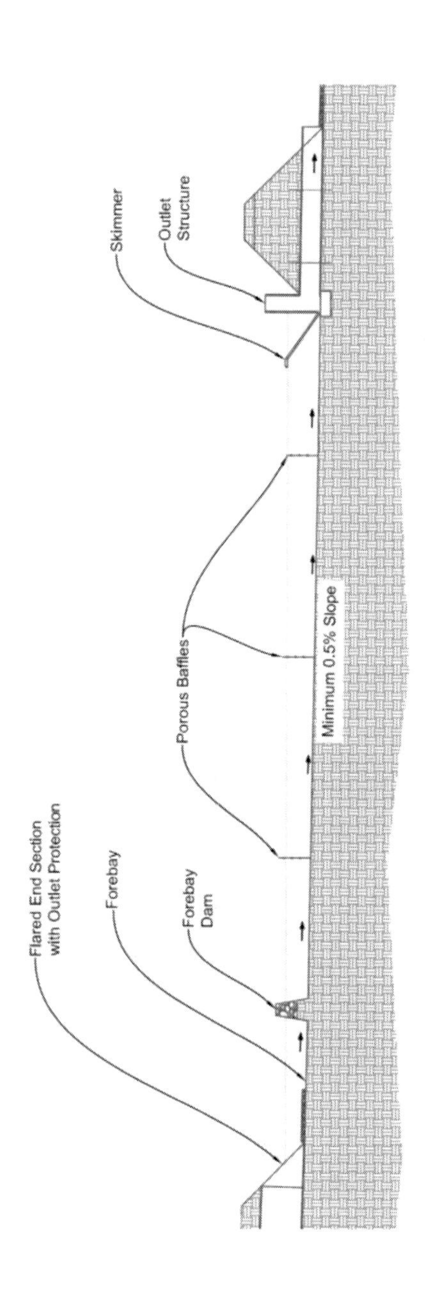

Silt Fence

Silt fence is typically placed downstream along the disturbed area to prevent sediment from sheet flowing off of the site. While silt fence does have some permeability, it is not meant as a filtration device. Rather, the efficacy of silt fence comes from its ability to hold stormwater and allow sediment to settle out prior to leaving the site. Based on DHEC regulations, there should be no more than ¼ acre of drainage per 100 linear feet of silt fence. If silt fence is placed at the bottom of steep slopes or along high flow paths, it will often overtop or fail.

Figure 5-3: Silt Fence

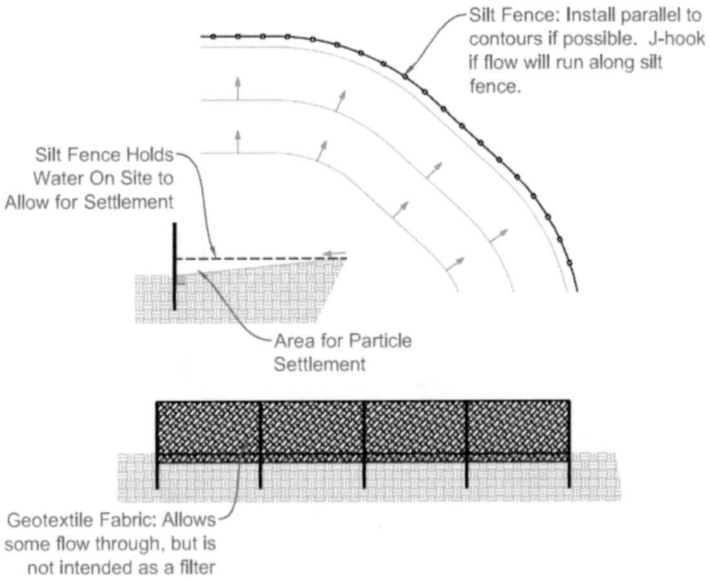

Check Dam

A check dam is used in swales to reduce flow velocity and capture sediment. Check dams can be designed with varying materials, but they are typically constructed with rocks or sediment tubes. Rock check dams are preferred for higher velocities or when the check dams must be utilized for a longer period of time.

Silt Fence Rock Outlet

A silt fence rock outlet is used when you have a discharge point along a stretch of silt fence. These will often be used at the discharge of a sediment basin to prevent the concentrated flow from overtopping or damaging the silt fence. Generally, a silt fence rock outlet is created by placing stones of varying sizes between an opening in the silt fence.

Inlet Protection

Inlet protection is placed on any inlets that may collect construction runoff. These can be inlets that are installed during construction, or existing inlets downstream of the runoff. There are many types of inlet protection for each various type of inlet. It is important to think conceptually about the purpose of the inlet protection when selecting the appropriate measure. For example, if you have high flows discharging to a yard inlet, simply placing silt fence around the inlet will not suffice during heavy storm events.

Construction Entrance

A construction entrance must be installed to prevent tracking when vehicles exit the construction site. The construction entrance is made of fabric and large stones that are meant to remove sediment as the construction vehicle leaves the site.

Pipe Slope Drain

A pipe slope drain is used in instances where there are steep slopes on a construction site that cannot be protected from erosion by other means. The slope drain is typically made of a small corrugated plastic pipe and facilitates the flow of stormwater from the top of the slope to the bottom, preventing heavy flows from eroding the slope. It is important to adjust the grading at the top of the slope to direct stormwater to the opening of the slope drain or it will not be effective.

Mulching

Mulching is a measure to prevent erosion that is caused by the impact of rainfall on exposed soil. Mulching is usually used in cases where an area will not be worked on for some time, and erosion is an issue.

Erosion Control Blanket

Erosion control blankets are used to help in the stabilization of slopes during and after construction. These will eventually biodegrade, so they will not be a permanent fixture.

Turf Reinforcement Mat

Turf reinforcement mats are used on slopes that are steeper than those that erosion control blankets would be used on. These are permanent, and care must be taken to insure they are installed properly, and grassing is completed, or they will become aesthetically displeasing and ineffective.

Outlet Protection

Outlet protection typically consists of rip-rap aprons placed at the end of a pipe outlet. In some cases, plunge pools are also used for outlet protection.

Design Process

Each project will have its own challenges when it comes to erosion and sediment control design. However, there will be some key components that are used for most projects. For example, sediment basins should be used in any case where there is more than ten acres of disturbance. Silt fence will be used in most cases where stormwater will sheet flow off the site. It is easiest to start with the more common BMP's and work your way through the design process.

One especially challenging aspect of erosion and sediment control design is the fact that the topography will change throughout project construction. For some projects, especially large ones, it may be necessary to have intermediate grading plans for proper conveyance of stormwater during construction activities.

I've only touched this topic briefly. The best way to learn about proper erosion and sediment control design is to visit construction sites. Nothing will be more beneficial to your understanding of the concepts and BMP's discussed in this chapter.

Six
Water

The Big Picture

While stormwater and wastewater are preferably conveyed via a gravity flow system, water is transported entirely through a pressurized system. To discharge from hydrants, sinks, or showers, the water must be under pressure within the system. This pressure can be conveyed in two primary ways, either through elevated tanks or through pumping systems. A water treatment facility is typically located such that it can take in flow from a nearby surface water source or a groundwater source. The treated water is then pumped into the distribution network.

You will notice elevated tanks at high points throughout a municipality. Water is pumped into the tanks and the resulting head provides pressure into the distribution system. These tanks are used for emergency and typical distribution depending on the design. Large water mains distribute water throughout the system, with branches reaching out to provide service to the houses, commercial facilities, and any other destination that requires a water source. Design of water distribution systems is based on providing water at adequate pressures. This includes the pressure necessary for basic use, such as showers, and for adequate pressures at hydrants and buildings for fire applications. This chapter will cover the conveyance of water primarily from a land development perspective. Therefore, special design component such as water tanks and booster pumps are not included.

Calculations

Basic Demand

Three primary demand categories for water are residential, commercial, and industrial. Each will have varying amounts of consumption, but there are standard guidelines for demand calculations provided by DHEC.

Residential

Residential demand refers to typical household usage such as showering, drinking, and cooking. Peaks typically occur during the morning between 7am and 11am and the evenings between 4pm and 8pm. Refer to **Figure 6-1** for a typical diurnal pattern showing the variation in water usage throughout the day.

Commercial

Commercial demand refers to a wide range of uses including department stores, office buildings, and restaurants. As you can imagine, each use will differ greatly in both the peak times as well as the average water usage.

Industrial

Industrial use can be the most difficult to quantify, there are general guidelines, but the designers should gather as much information about the proposed use to determine proper water demands.

Reference can be made to **Table 6-1** for some common demand values provided by DHEC. A complete list can be found in **Appendix D.** These are wastewater loading guidelines, but often the demand used in water is the same as the loading used in wastewater calculations. Caution should be taken with this approach. If you think about it, it makes some sense that the quantity of water that leaves the faucet, shower, or toilet will match the quantity that enters the sewer system. However, this does not account for water used for outdoor purposes that doesn't enter the sewer system. In dry regions, a significant amount of the household water usage may be utilized for outdoor purposes.

Table 6-1: Typical Demand Values[13]

Type of Establishment	Hydraulic Demand (GPD)
B. Apartments, Condominiums, Patio Homes:	
1. Three (3) Bedrooms (Per Unit)	300
2. Two (2) Bedrooms (Per Unit)	225
3. One (1) Bedroom (Per Unit)	150
L. Clinics, Doctor's Office:	
1. Per Employee	11
2. Per Patient	4
N. Dentist Office:	
1. Per Employee	11
2. Per Chair	6
3. Per Suction Unit; Standard Unit	278
4. Per Suction Unit; Recycling Unit	71
5. Per Suction Unit; Air Generated Unit	0
O. Factories, Industries:	
1. Per Employee	19
2. Per Employee, with Showers	26
3. Per Employee, with Kitchen	30

[13] (South Carolina Department of Health and Environmental Control, 2002)

Type of Establishment	Hydraulic Demand (GPD)
4. Per Employee, with Showers and Kitchen	34
Q. Grocery Stores: (Per Person, No Restaurant or Food Preparation)	19
R. Hospitals:	
1. Per Resident Staff	75
2. Per Bed	150
S. Hotels: (Per Bedroom, No Restaurant)	75
Z. Offices, Small Stores, Business, Administration Buildings: (Per Person, No Restaurant)	19
CC. Residences: (Per House, Unit)	300
FF. Restaurants:	
1. Fast Food Type, Not Twenty-Four (24) Hours (Per Seat)	30
2. Twenty-Four (24) Hour Restaurant (Per Seat)	53
3. Drive-In (Per Car Service Space)	30
4. Vending Machine, Walk-up Deli or Food Preparation (Per Person)	30
GG. Schools, Day Care:	
1. Per Person	8
2. Per Person, with Cafeteria	11
3. Per Person, with Cafeteria, Gym and Showers	15
II. Shopping Centers, Large Department Stores, Malls: (Per Person, No Restaurant)	19

Figure 6-1: Typical Residential Diurnal Pattern

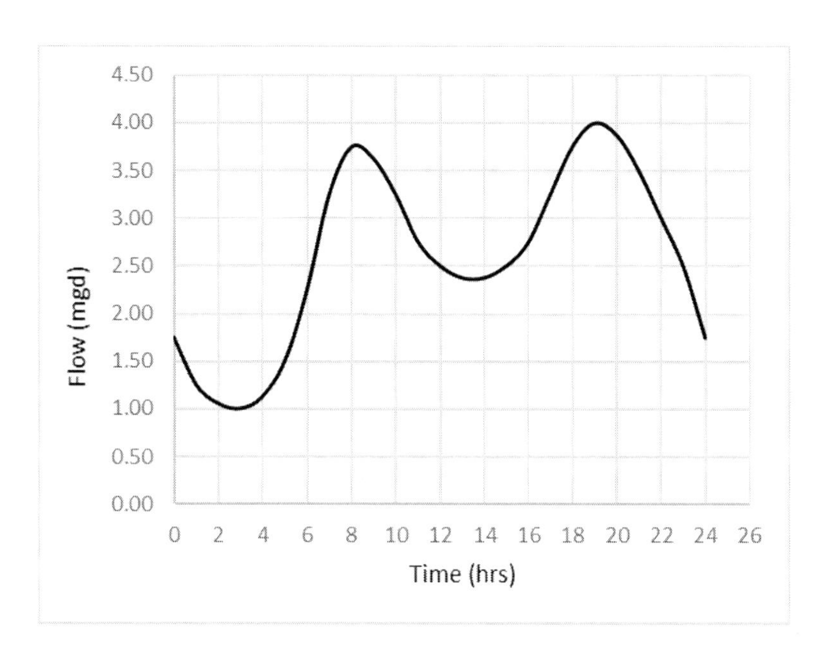

Peaking Factors

Typical water demand values are usually provided in gallons per day (gpd). However, as was previously discussed, water usage varies greatly throughout the day. So, if a residence uses 300 gallons per day, this does not mean that they are using this flow at a constant rate throughout the day. Rather they may produce a demand of 600 gpd in the morning, but 0 gpd at 3 am. A typical diurnal pattern can be seen in **Figure 6-1**. For simplified calculations, a peaking factor of 2.5 is common. Following is the peak flow calculation for a single-family residence:

$$Peak\ Hourly\ Flow = Peaking\ Factor * Average\ Daily\ Flow$$

$$Peak\ Hourly\ Flow = 2.5 * 300 = 750\ gpd$$

Design Elements

Pipes

Pipe Material

The pipe material used will typically depend on the required size and pressure that will be applied to the pipe. The most common pipe materials used for water system design are DIP, PVC, HDPE, and steel. In high pressure areas, DIP will be necessary as PVC cannot withstand high pressures. DIP will also be used when minimum cover requirements cannot be met or when minimum separation requirements cannot be met. There is obviously a cost consideration when selecting materials, and PVC is the least expensive option if it will meet the design requirements.

HDPE is primarily used in directional drilling applications. This material allows for jointless connections that can facilitate curved alignments without trenching. This will be used in cases when it is necessary to traverse long distances under wetlands or streams where disturbance is not allowed.

Steel pipe is only used in very large applications, typically near the water treatment plant. **Table 6-2** provides a general overview of pipe materials used for water system design.

Table 6-2: Water System Pipe Materials

Material	Comments
Polyvinyl Chloride (PVC)	Used in Small Systems
Ductile Iron Pipe (DIP)	Large Systems. High Pressure Areas. Insufficient Cover. Insufficient Vertical Separation.
HDPE	Directional Drilling
Steel Pipe	Large Water Mains. Near Treatment Facility.

Pipe Sizing

Pipe sizing is dependent on pressure requirements. If a certain pipe size, such as a 6-inch pipe, is not providing adequate pressure, you should increase the size to an 8-inch and determine if this is adequate. The following section contains details on why this is the case. Six inches is typically the minimum size for a main as this is the size required for a standard hydrant. It is possible to use a 4-inch pipe in some areas if a post hydrant is use.

In Depth: Effect of Pipe Size on Pressure

The most common misconception in water distribution system design is that a reduction in pipe size will result in an increase in pressure. It is understandable that intuition would lead to this misunderstanding. Hopefully, what follows will clear things up and bring you to the realization that this phenomenon does make intuitive sense.

First, let's see what Bernoulli's equation tells us about pressure changes in pipe systems. As you might recall, Bernoulli found that if we make some basic assumptions regarding flow conditions (laminar, steady flow), we can use the principle of Conservation of Energy to find the relationship between elevation, velocity, and pressure in a pipe at any given point.

Bernoulli's Equation

$$P_1 + \frac{1}{2}\rho v_1^2 + \rho g h_1 = P_2 + \frac{1}{2}\rho v_2^2 + \rho g h_2$$

$$P = Pressure\ Energy;$$

$$\frac{1}{2}\rho v^2 = Kinetic\ Energy;$$

$$\rho g h = Potential\ Energy$$

- *P: Pressure*
- *ρ: Density*
- *v: Velocity*
- *h: height*

We also know the principle of continuity indicates that the volumetric flow rate will remain the same; $Q_1=Q_2$. Therefore, $V_1A_1=V_2A_2$. This is how nozzles work. This is also where the confusion begins for many people. We learn early on that we can increase the velocity of water from a hose by placing our thumb over part of the end. We also know that if we hold our hand in front of the water exiting the hose, it will feel like it has more pressure when we are constricting the flow path. In order to clear up this misunderstanding, we need to look a little closer at the type of pressure Bernoulli's equation is referring to.

The pressure your hand feels when holding it at the end of a hose is created by the rate of change of momentum as the velocity of the water is abruptly halted. The pressure we are analyzing in pipe systems is the pressure that is exerted by the fluid on itself and the pipe walls. It is easier to understand this concept by looking at how this pressure is measured using piezometers. As we can see in **Figure 6-2**, at point 1, the pressure as measured by the head in the piezometer is higher than the pressure at point 2.

Now let's imagine the water is exiting the pipe and we place another piezometer at the edge of the stream of water (Point 3). We will also imagine the water will be hitting a wall shortly after exiting the pipe. Just thinking intuitively, do you think any water would actually fill the piezometer? The answer is no, since the pressure that would generate this head is essentially zero after the water exits the pipe. Now if we look at the force on the wall, we see that this is caused by the rate of change of momentum, which is clearly different from the type of pressure that would fill the piezometer.

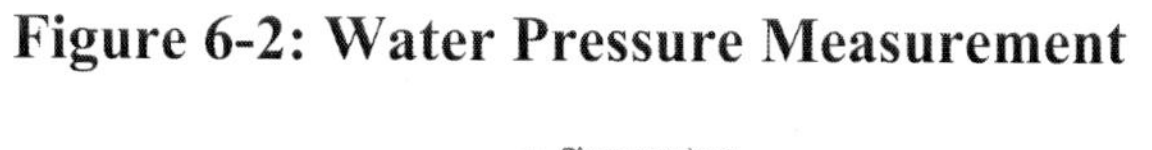

Figure 6-2: Water Pressure Measurement

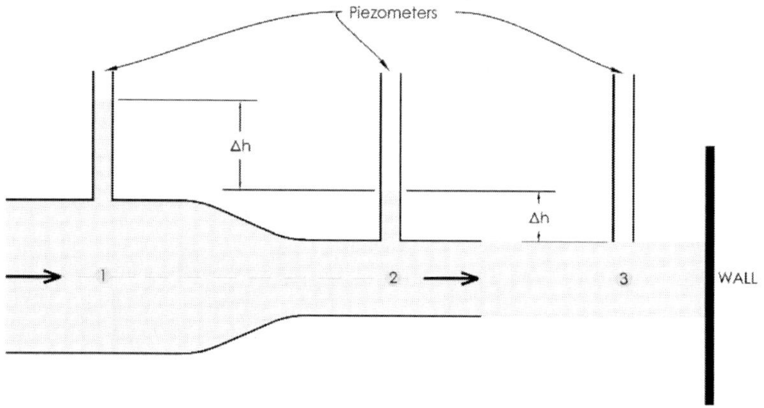

Hopefully this clears up what type of pressure we are analyzing in a pipe system, but it doesn't explain why the internal pressure is less at point 2 than it is at point 1. After all, aren't we compressing the fluid which would increase the pressure on the pipe walls? Not exactly. Water is incompressible. This is why the decrease in area must result in an increase in velocity. Since the flow rate must remain the same, water in the smaller pipe must move faster in order to move the same differential volume over a longer distance (dL) in the same amount of time. This can be seen more clearly in **Figure 6-3**.

In order to understand how pressure changes with velocity, it may be best to think of the fluid on a molecular level. Remember, we are talking about the pressure on the pipe walls. This pressure is the result of molecules hitting the sides of the pipe. Therefore, the motion of the molecules is what generates this pressure. As the velocity in the pipe decreases, the motion of the molecules is more random, meaning

Figure 6-3: Differential Volume

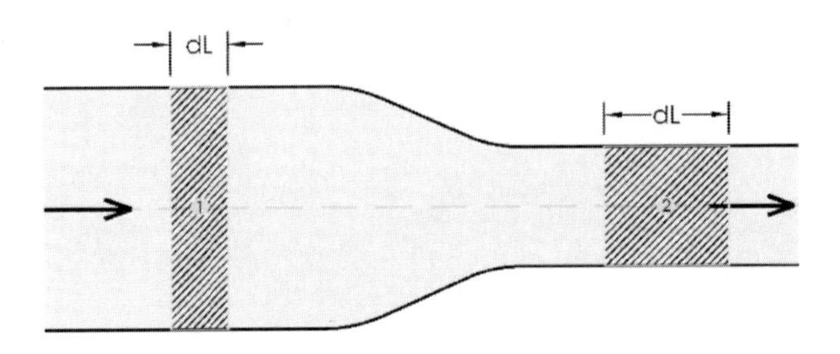

there are more collisions with other molecules and the pipe walls. As the velocity increases, the molecules must move in a more ordered manner, resulting in the molecules moving more toward the direction of flow and reducing collisions, thus decreasing pressure.

So this explains why a smaller pipe size will result in lower pressures, but if you look back at Bernoulli's equation, you'll see that we haven't fully answered the question as it pertains to a water distribution system. Imagine we have a system that has a smaller pipe in between two larger ones (See **Figure 6-4**). If we apply the equation we used earlier, we will see that the pressure at point 3 will be the same as the pressure at point 1. If this were actually the case, we could design systems that only had large pipe diameters in areas where there was a demand.

Figure 6-4: Changing Pipe Sizes

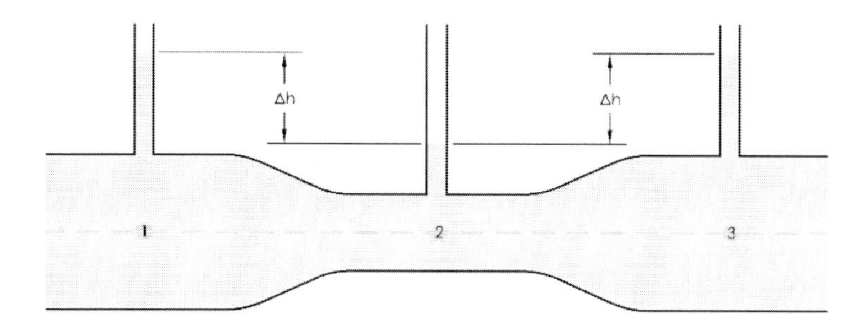

That is because there is one important aspect of Bernoulli's principle that is typically left out in basic flow equations; head loss. With head loss added to the equation, we get this:

$$P_1 + \frac{1}{2}\rho v_1^2 + \rho g h_1 = P_2 + \frac{1}{2}\rho v_2^2 + \rho g h_2 + \sum h_l$$

As the velocity in a system increases, the head loss also increases. Since a smaller pipe size results in higher velocities, the total head loss in the system is greater; meaning there is less available pressure energy.

Now that we've looked at a lot of scenarios and equations, let's briefly summarize the overall concept. We already knew that, based on Bernoulli's equation, decreasing pipe size also decreases pressure. Now we also know that the increased velocities in smaller pipes result in greater friction losses in the system. So, increasing the pipe size results in greater available pressures and a decrease in the overall friction loss in the system.

Minimum Cover

The standard minimum cover requirement for a water main is 3 feet. This may vary depending on the loading conditions above the pipe and the potential for freezing depending on the local climate. During the design process, it is important to give yourself more than 3 feet of cover to allow for variations in the grade over the pipe and construction tolerance. If you design a pipe with exactly 3 feet of cover, it is possible that the installed pipe will have less than 3 feet of cover in some areas and you will not receive the necessary permit to operate until this is corrected.

Minimum Separation

The standard minimum vertical separation is 18 inches. It is preferred that the water main is installed above any wastewater mains to prevent contamination. It is also necessary to provide a minimum of 10 feet of horizontal separation between the water main and any proposed or existing wastewater main. In cases where the 18 inches of vertical separation cannot be met, you should coordinate with the regulatory agency to determine methods of pipe encasement that will allow a reduction of this requirement. Caution should also be taken when you have large storm drainage pipes above your water main. In these cases, you may need to increase the separation or use DIP to prevent damage from the vertical load.

Easements

Easements will be required in any case where the public entity will maintain ownership and maintenance responsibilities of the water main. A typical easement width is fifteen feet, but this will vary depending on agency guidelines, depth, and size. The main will need to be accessible in any case where maintenance is needed, and an adequate easement is needed to allow for this maintenance.

Anti-Seep Collar

Anti-seep collars are used whenever there is a potential for the seepage of water to undermine the interface between the pipe and the soil. The collars are designed to increase the distance that the water must travel along the pipe. There is not a lot of consensus regarding the efficacy of this method, but it is still commonly used in design. These are often used in areas where there is a high groundwater table or saturated soils such as wetlands or stream crossings.

Joints

Now that we have the necessary information to select our pipe materials, we must look at how the pipes are connected to put together our distribution system. Below is a basic list and description of common joints used in a water distribution system:

Push-on Joint

The push-on joint is easily installed and the least expensive option. However, there is no restraint provided by the push-on joint, and if used incorrectly, this joint can easily separate as a result of water hammer. The use of joint restraint and thrust blocking to prevent this separation will be discussed shortly.

Mechanical Joint

Mechanical joints provide more restraint than a push-on joint but are not considered fully restrained unless a restraining collar is used.

Flanged Joint

The flanged joint is a restrained joint but is expensive and installation is difficult. For this reason, the flanged joint is only used in cases where the cost is justified. These will most commonly be used at pump stations.

Joint Restraint

In any case where bends or valves will result in a force that might cause joint separation, restraints are used to prevent this. It is common to use both thrust blocking and joint restraints to prevent pipe separation, though there is little consensus on whether it is necessary to use both. The concepts of thrust blocking and joint restraint are shown in **Figure 6-7**.

Fittings

Now that we know how to put the pipes together, we need some way to change directions. You will often see design drawings that show a curved water main. Keep in mind that this curve effect is usually created by straight segments of pipe. For curves with a large radius, a simple deflection at the joint can provide the bend necessary to curve the alignment. This allowable deflection is usually somewhere between 2 and 5 degrees but is dependent on the pipe material and type of joint used.

When the pipe must make significant changes in direction or curve along a small radius, then fittings will be needed to allow for this. Fittings will also be used to allow sections of pipe to cross each other or for one segment to branch off of another. Following is an overview of common fittings used in water system design. These fittings are illustrated in **Figure 6-6**.

Figure 6-5: Types of Water Main Joints

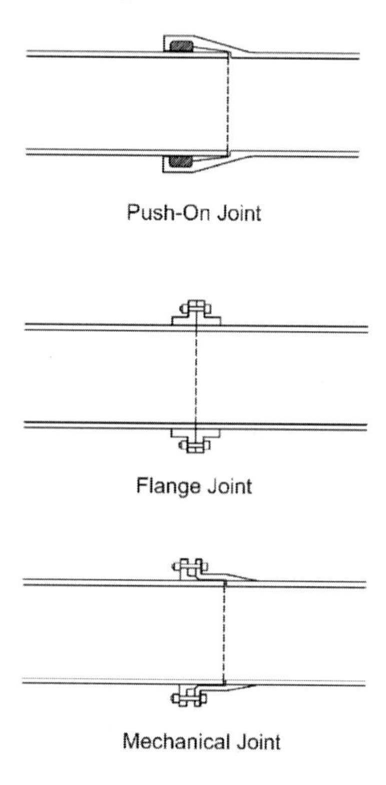

Bends

Bends allow for a change in direction, both horizontal and vertical. It is important to understand during your design process that these bends are not available in any size you want. There are specific sizes available, and you must ensure that your alignment properly accounts for this. Bends are typically available in thc following sizes: 11.25 degrees, 22.25 degrees, 45 degrees, and 90 degrees. You must use some combination of these fittings (along with small amounts of deflection) to properly layout your alignment. This may take some creativity and practice.

Tees

Tees are used when you want to branch off of a main line. For example, if you want to install a hydrant off of a 12" water main, you will need to add a 12" x 6" Tee to facilitate this. It is not typical to use a Tee for any size reduction along the main. In other words, the "top" of the tee will be the same size (12" in our example) and the branch portion can be a different size.

Cross

A cross, as the name implies, is used to cross water mains. Like the tee, these are not typically used for a size reduction along a water line. In other words, the pipe size will be the same on opposite sides of the cross.

Reducer

A reducer is used to change pipe sizes along a water main. Reducers are straight segments. If a pipe reduction is warranted near a bend or cross, it is preferable to place it before the fitting if possible, for the obvious cost savings.

Thrust Blocking

When water is flowing through a pipe at high pressures and encounters a bend, valve, or other obstruction, it will create a force that can pull the pipe apart at the joint. For this reason, thrust blocking is used at locations where this force is expected. Thrust blocking usually consists of concrete placed against compacted soil to prevent the fitting from moving when the water pressures exert a force on the fitting.

Figure 6-6: Water Main Fittings

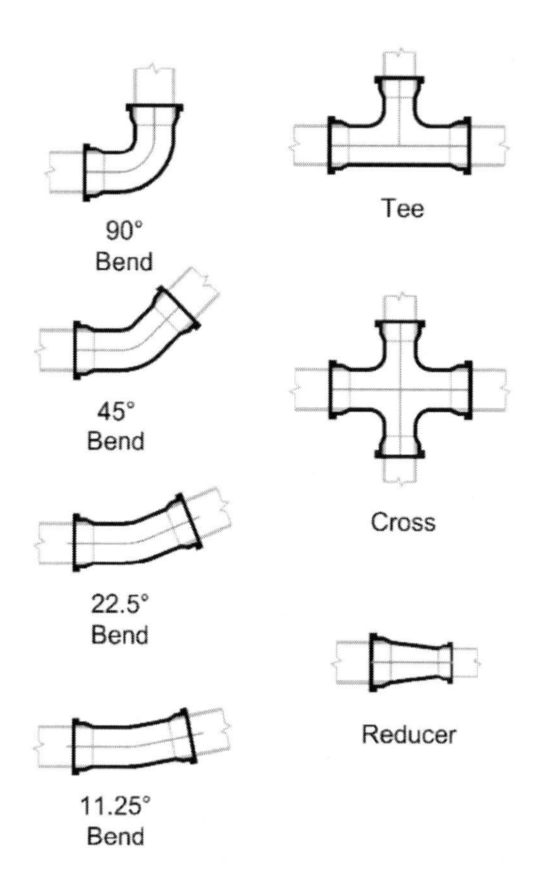

Figure 6-7: Thrust Blocking and Joint Restraint

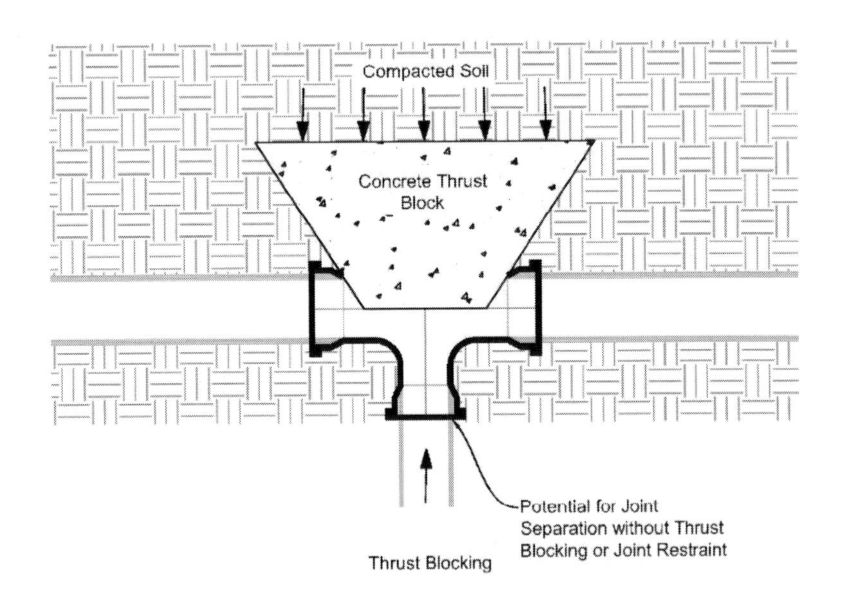

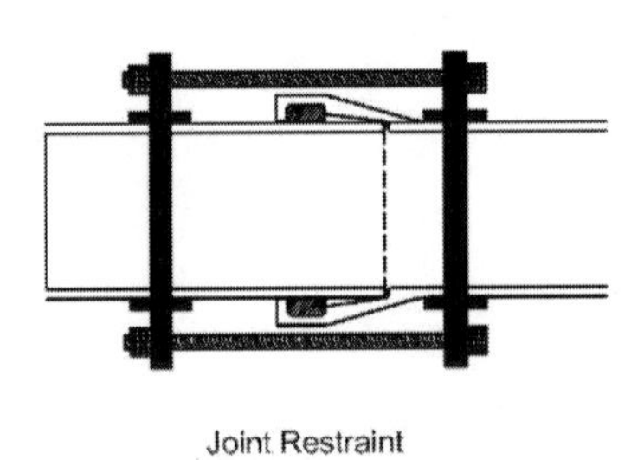

Joint Restraint

Valves

Gate Valve

This type of valve is used to isolate flow and is sometimes called a shutoff valve. Gate valves are designed to be fully opened or fully closed. They should not be used to partially restrict flow.

Pressure Reducing Valve

This type of valve is automated and reduces pressure to prevent damage to water lines and appurtenances when there is excess pressure. These can be placed on main lines as wells as small household service lines.

Check Valve

This type of valve creates boundaries in the system by preventing backflow.

Blowoff Valve

This type of valve is used to dewater the line at dead ends and low points to prevent stagnant water from building up in the system.

Air-Release Valve

The type of valve releases air that builds up in a water main. Dissolved air within the water builds up in turbulent flow conditions and rises to the high points along a water main. If this air is not released it can severely inhibit flow, sometimes stopping it completely. Air release valves are typically installed at high points and at any change in slope along the main.

Figure 6-8: Common Water Valves

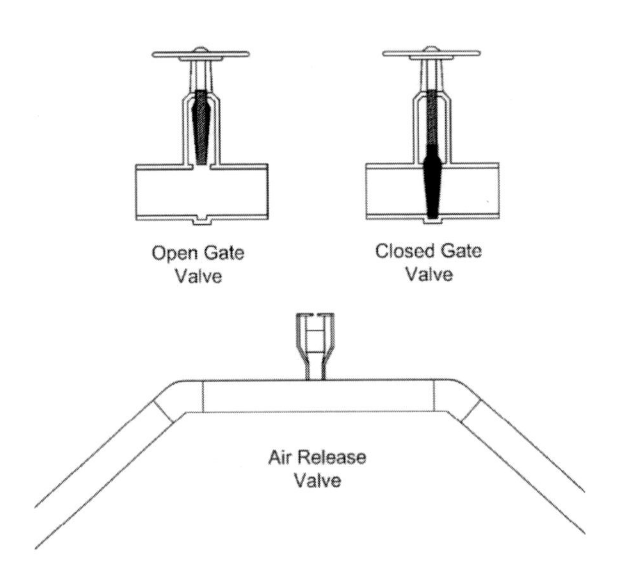

Appurtenances

Fire Hydrant

A typical fire hydrant requires a minimum 6-inch water main for connection. Hydrants usually have two 2 ½-inch hose connections and one 4 ½-inch pumper connection. Thrust blocking is always used during installation to prevent pipe separation due to changing pressures. Measures should also be taken to prevent settling and allow for proper drainage of water around the base of the hydrant. An illustration of a typical hydrant can be found in **Figure 6-9**.

Figure 6-9: Standard Fire Hydrant

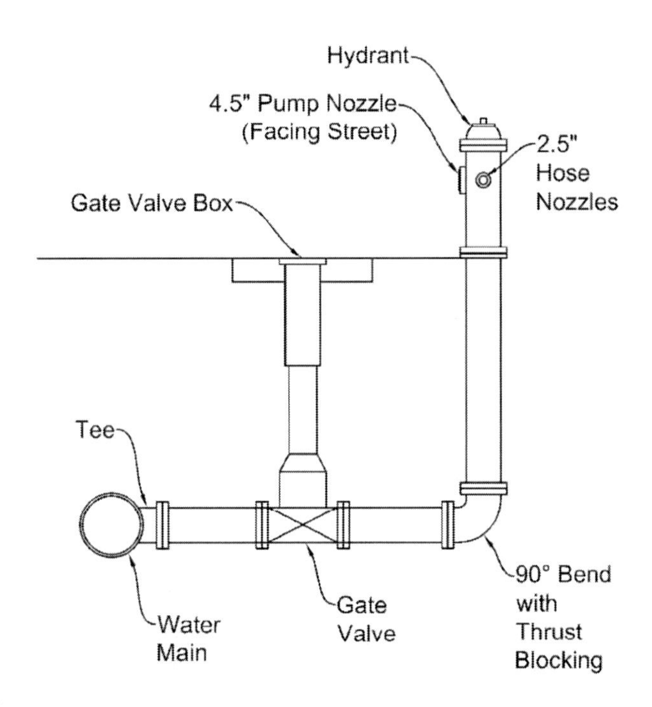

Service Connection

Service connections are needed to connect the water source with its destination. Each house or business has a service connection to the water main that allows for distribution of the water as well as metering for billing purposes. For smaller buildings, such as residences, a small meter box is used to house the domestic service meter.

Vault

In commercial, industrial, or multi-family structures, it will be necessary to supply a fire line to the building in addition to a domestic service line. These structures also require backflow preventers. The meter for the fire line and the backflow preventer (usually dual check valves) are housed in a large vault. This vault is typically made of precast concrete and is approximately 6' x 10' in size.

Installation Techniques

To assist with the design process, you will need to understand some installation techniques. Below are the common installation techniques you will use in water main design. All of these, except the horizontal directional drill, are applicable to stormwater and wastewater system installation as well.

Trenching

Piping is laid in trenches that are usually bedded with compacted granular material. The contractor should take great care to level out the bedding, so the pipe isn't impacted by varying loads that will cause deflection or fractures. The mode of pipe failure depends on the type of pipe. Flexible pipes such as plastic and ductile iron will deform, while rigid pipes such as concrete will fracture.

Open Cut

Open cut is a method used when it is necessary to install a pipe across a pavement section. The contractor will sawcut the asphalt, install the pipe and backfill, then replace the pavement section. It is also common for municipalities to require that a certain amount of pavement surrounding the open cut area is milled and overlaid to allow for a smooth transition.

Jack and Bore

In many cases, it is more cost effective or required by the municipalities that the pipe be installed via jack and bore. This technique involved the installation of a steel casing by drilling horizontally under the roadway. The pipe is then installed inside the steel casing. There will be a bore pit and receiving pit on the sides of the road, but it is not necessary to disturb the actual roadway. This technique is not practical for long distances and does not allow for any redirection between the bore pit and receiving pit.

Directional Drill

Directional drill is typically used when you must install the water main under a wetland or stream. Unlike jack and bore, it does not require deep pits to be dug, and it allows for a change of direction between the start and end. For this application, a hole is drilled, usually curving under the wetland, stream, or other obstacle and the pipe is pulled back through. It is relatively easy to install the pipe very deep this way, but you should make sure proper geotechnical exploration is completed so that you are aware of the material that the drill will encounter. HDPE pipe is used with directional drill.

Figure 6-10: Trenchless Installation Techniques

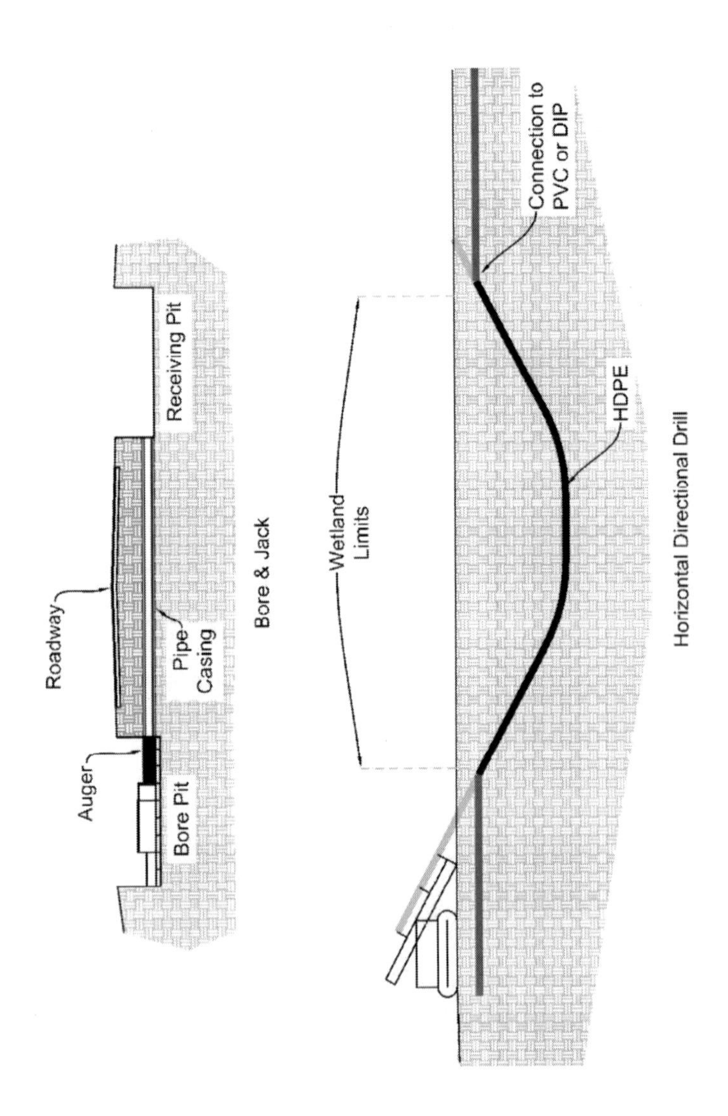

Design Process

When designing a new neighborhood or commercial development, it is necessary to analyze whether the existing water distribution network will provide adequate pressure. This pressure must be sufficient to provide for basic necessities such as showering and must also provide adequate pressure for firefighting purposes. Hydrant pressure requirements vary, but a typical requirement is a minimum of 25 psi of residual pressure. Fire trucks have pumps, but if there is not enough pressure at the hydrant they are pulling water from, the water main could collapse due to negative pressures.

Looping is also an important concept in design. If there are long straight dead-end section so pipe, the water will become stagnant. In cases where there are dead ends, hydrants or blow off valves should be used to allow the pipe to be cleaned. In this case, calculations may be needed to ensure there is adequate cleaning velocity when the hydrant or valve is opened.

Water main design is relatively straightforward if you understand the general goals described above (it gets a bit more complex for larger projects). There are certainly some specifics, such as valve and hydrant placement, that are more complicated, but you will be able to grasp these design parameters easily if you understand the overall concepts described in this chapter. Some examples of water main plan and profile design concepts are shown in the figures at the end of the chapter.

Figure 6-11: Water Main Plan (Commercial)

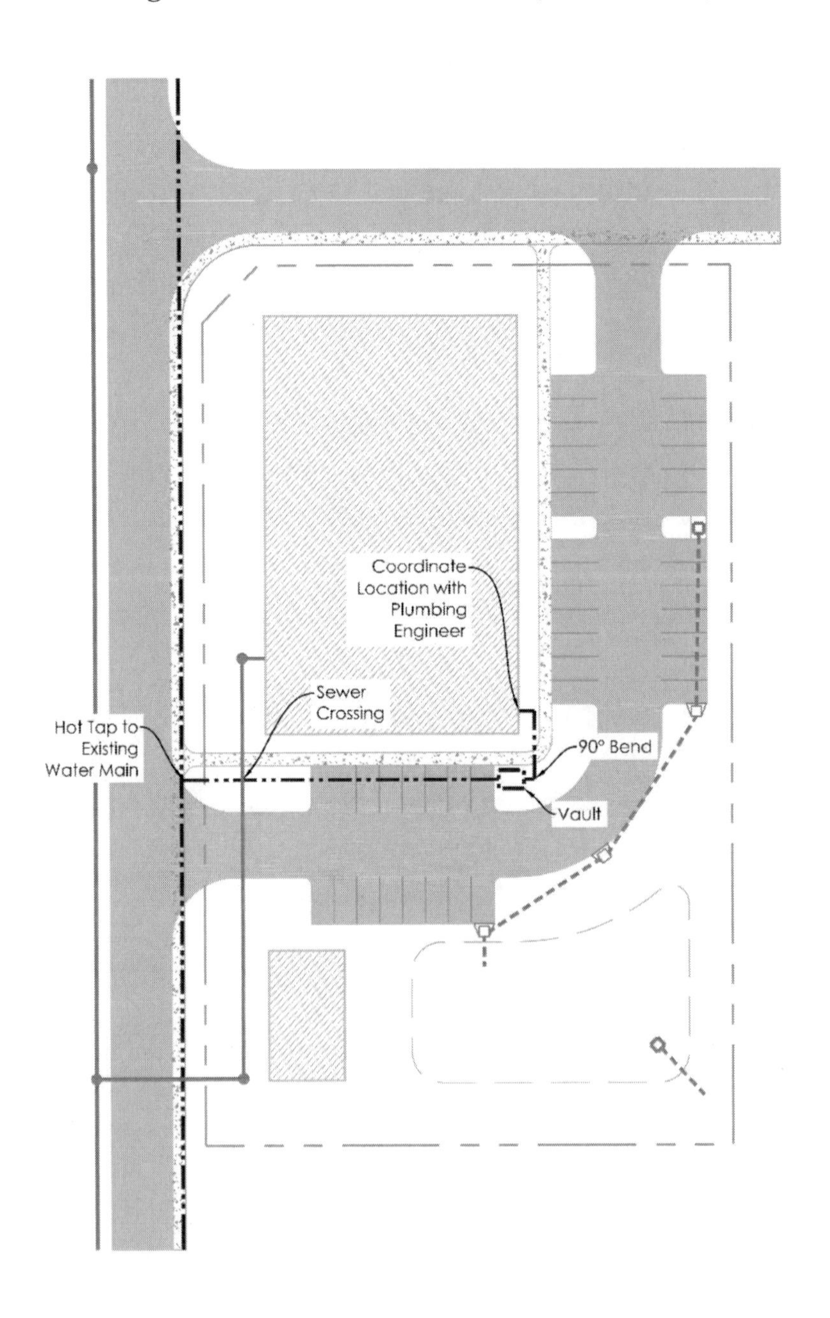

Figure 6-12: Water Main Plan (Residential)

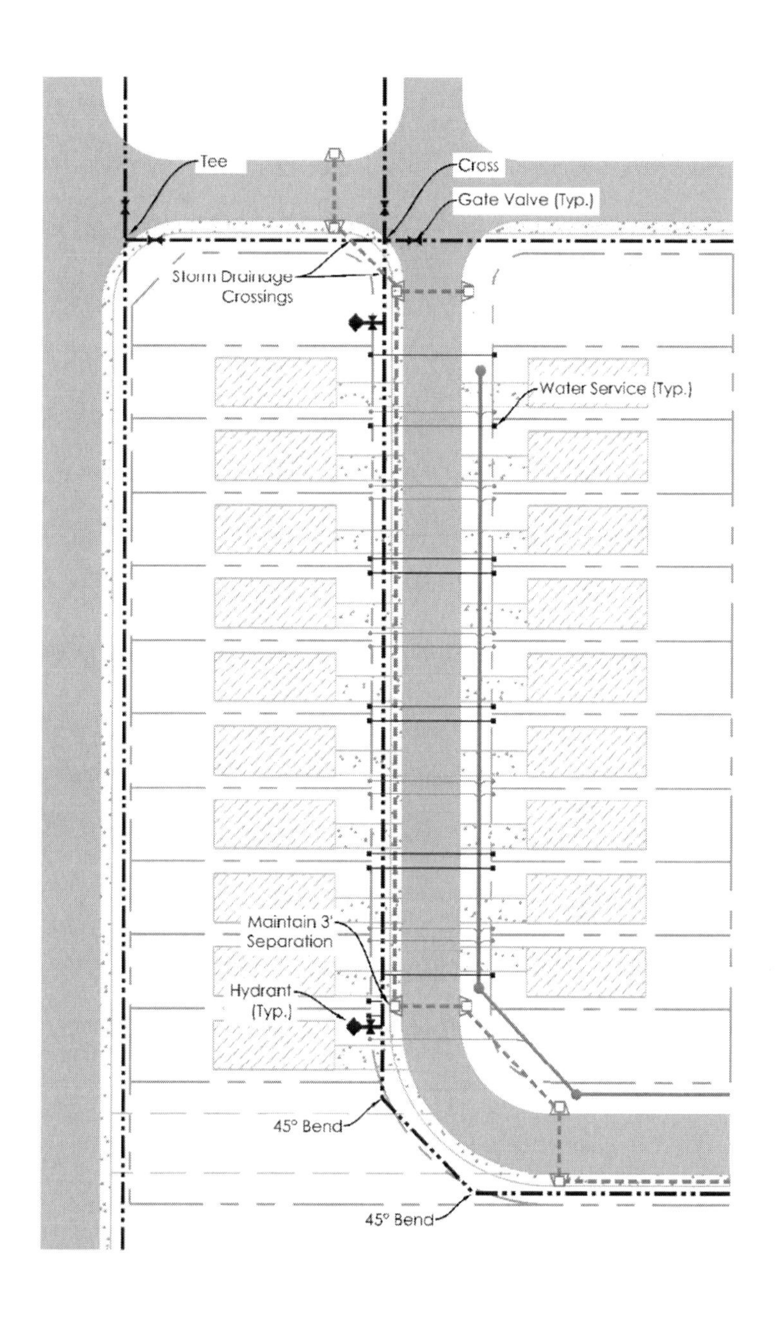

Figure 6-13: Water Main Profile

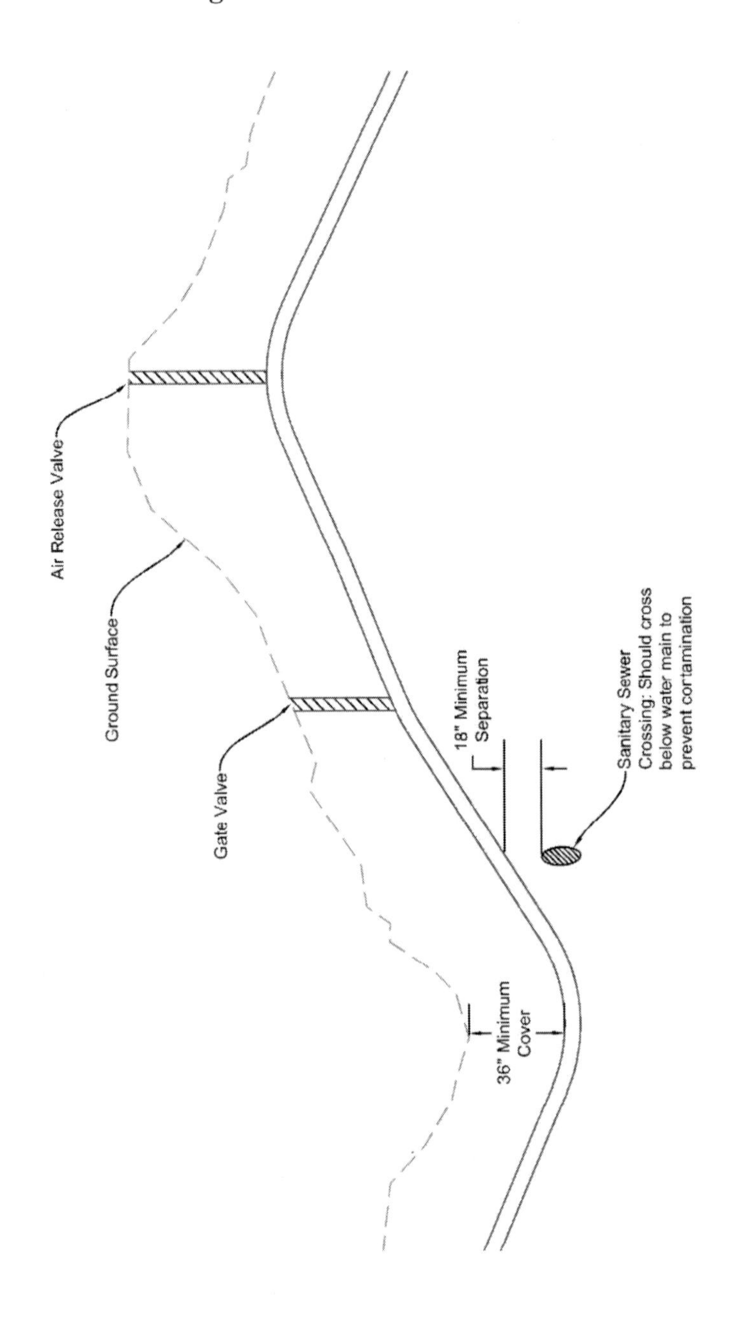

Seven
Wastewater

The Big Picture

The previous chapter discussed how water gets to your house. Now consider how the water that exits. Where does it go after it enters the drain? It is conveyed through a series of small pipes until it exits your house. Then it will discharge to a sewer collection system consisting of pipes and pump stations that convey the wastewater to a treatment plant. At the treatment plant, it goes through a very thorough treatment process before it is discharged back to the environment. This text will cover the point after it exits your house to just before it enters the treatment plant.

Conveyance and treatment of wastewater is vital to a healthy environment. Wealthy countries like the United States are fortunate to have advanced systems that have all but eliminated water borne illness. However, many people in less-fortunate societies suffer the consequences of inadequate and non-existent wastewater collection systems. This is something that we should not take for granted.

Wastewater is ideally conveyed by gravity. This is cheaper and easier to maintain. Routes are selected based on the topography and existing land use - to provide maximum service to an area for minimal costs. Major gravity sewer systems often follow streams and rivers as these are naturally the low spots in a sewer basin. There are many instances where gravity systems will not be feasible, and wastewater must be transported against the energy gradient. In these cases, pump stations are used. Pump stations add energy to the system to move the wastewater to higher elevations.

Calculations

To properly design a wastewater conveyance system, you must have some understanding of how much wastewater it will need to convey. For wastewater, this is called loading. There are standard values that are provided by DHEC for various uses which will be discussed in the next section. This section will focus on loading calculations. Calculations for pipe sizing and pump station design are included in the design elements section.

Base Loading

The primary loading categories for wastewater are residential, commercial, and industrial. Each will have varying amounts of loading, but there are standard guidelines. Below are some standard loadings for common land uses. The full list is provided in **Appendix D**.

Table 7-1: Typical Demand Values

Type of Establishment	Hydraulic Loading (GPD)
B. Apartments, Condominiums, Patio Homes:	
1. Three (3) Bedrooms (Per Unit)	300
2. Two (2) Bedrooms (Per Unit)	225
3. One (1) Bedroom (Per Unit)	150
L. Clinics, Doctor's Office:	
1. Per Employee	11
2. Per Patient	4

Type of Establishment	Hydraulic Loading (GPD)
N. Dentist Office:	
1. Per Employee	11
2. Per Chair	6
3. Per Suction Unit; Standard Unit	278
4. Per Suction Unit; Recycling Unit	71
5. Per Suction Unit; Air Generated Unit	0
O. Factories, Industries:	
1. Per Employee	19
2. Per Employee, with Showers	26
3. Per Employee, with Kitchen	30
4. Per Employee, with Showers and Kitchen	34
Q. Grocery Stores: (Per Person, No Restaurant or Food Preparation)	19
R. Hospitals:	
1. Per Resident Staff	75
2. Per Bed	150
S. Hotels: (Per Bedroom, No Restaurant)	75
Z. Offices, Small Stores, Business, Administration Buildings: (Per Person, No Restaurant)	19
CC. Residences: (Per House, Unit)	300
FF. Restaurants:	
1. Fast Food Type, Not Twenty-Four (24) Hours (Per Seat)	30
2. Twenty-Four (24) Hour Restaurant (Per Seat)	53
3. Drive-In (Per Car Service Space)	30
4. Vending Machine, Walk-up Deli or Food Preparation (Per Person)	30
GG. Schools, Day Care:	
1. Per Person	8
2. Per Person, with Cafeteria	11
3. Per Person, with Cafeteria, Gym and Showers	15
II. Shopping Centers, Large Department Stores, Malls: (Per Person, No Restaurant)	19

If this seems familiar, that's because it is identical to the water base demand discussed in the previous chapter. In most cases, the water demand is assumed to be the same as the wastewater loading even though this is not always true. In drier climates, where excessive water use for irrigation is expected, care should be taken to factor this into the water demand calculations.

Residential

Residential loading refers to typical household usage like showering, washing dishes, and doing laundry. Peaks typically occur during the morning between 7 am and 11 am and the evenings between 4 pm and 8 pm.

Commercial

Commercial demand refers to a wide range of uses including department stores, office buildings, and restaurants. Each use will differ greatly in both the peak times as well as the average loading.

Industrial

Industrial use can be the most difficult to quantify. There are general guidelines, but the designer should gather as much information about the proposed use to determine proper loading.

GWI and RDI/I

In addition to the base loading discussed previously, sewer systems are susceptible to groundwater infiltration (GWI) and rainfall-derived inflow and infiltration (RDI/I). These concepts are discussed in further detail in the wastewater system analysis section.

Figure 7-1: Typical Residential Diurnal Pattern

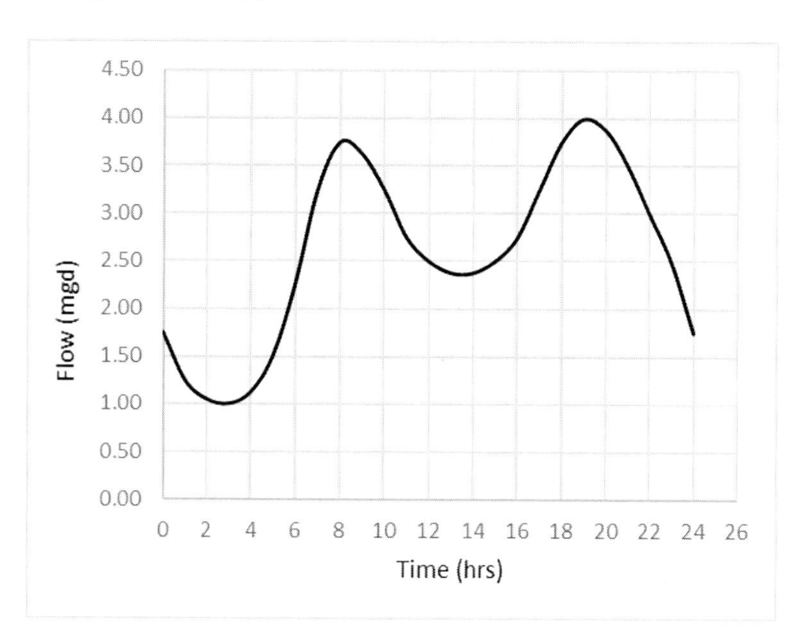

Peaking Factors

Loading values give us the average flow, but this flow will not occur at a steady pace throughout the day. There will be peaks during morning and evening hours, and downtimes during midday and midnight hours. Therefore, peaking factors are needed to adequately size infrastructure. A peak factor of 2.5 is typical for system analysis. Refer to the diurnal pattern in **Figure 7-1** for a better understanding of the fluctuation of wastewater loading throughout a typical day for a residential property.

In Depth: Wastewater System Analysis

For system-wide wastewater system planning, it is often necessary to perform an analysis of the system and develop a calibrated system model. Following is an overview of the typical process for this analysis.

Data Collection

Three sets of data are needed to complete an accurate analysis of the existing wastewater collection system: physical system data, flow monitoring data, and rainfall data.

Physical System Data

Physical system data is required to build a computer model of the collection system for analysis. The following data components are needed:

- *Manholes*
    - *Location*
    - *Rim Elevation*
    - *Invert Elevation*
    - *Diameter*
- *Gravity Mains*
    - *Location*
    - *Invert Elevations*
    - *Diameter*
    - *Material*
- *Pump Stations*
    - *Location*
    - *Wet Well Parameters*
    - *Pump Information*
    - *Control Settings*
    - *SCADA Information*
- *Force Mains*
    - *Location*
    - *Profile Elevations*
    - *Diameter*
    - *Material*
    - *Valve Locations*

Flow Monitoring Data

Flow monitoring data is used for two purposes: the analysis of the GWI and RDI/I contributions to the system and the calibration of the sewer model. To collect flow monitoring data, meters should be placed strategically throughout the system to provide good spatial coverage and capture flow that will provide a good understanding of the systems wet-weather response.

Rainfall Data

Rainfall data is necessary to characterize the wet weather response of the system. This information will be used to determine the RDI/I contribution and determine the parameters necessary to model the response to a specified storm event.

Sub-Basin Delineation

Once you have a system layout and meter locations, you can delineate the sub-basins. Sub-basins can be delineated using various methods, but when you have flow meter data, it is recommended that they be delineated based on these flow meter locations. The concept is the same as the delineation of a watershed; you will determine the outfall (meter location) and then identify all pipes and pump stations contributing flow to that outfall.

Flow Monitoring Analysis

An analysis of the flow monitoring data helps to determine areas that are susceptible to RDI/I. Following is an overview of how the flow monitoring data is processed.

Flow Components

As I mentioned earlier, there are three flow components that contribute to wastewater collection and conveyance: BWWF, GWI, and RDI/I. These three components are illustrated in **Figure 7-2**.

BWWF

This is the base loading that comes from the residential, commercial, or industrial development. BWWF is what is expected to enter the system through proper connections. The BWWF component is identified in **Figure 7-2**. Notice that this typically follows the diurnal pattern we discussed previously, assuming a majority of the loading sources are residential.

GWI

Wastewater collection systems are not watertight. Defects in the manholes and pipes allow for the infiltration of groundwater into the system. This infiltration is relatively steady and constant but does fluctuate seasonally as the groundwater levels vary. This flow component is not impacted by rainfall on a short-term basis. See **Figure 7-2** for a representation of this flow.

RDI/I

There are two components to RDI/I: inflow and infiltration. Inflow is stormwater that immediately enters the system, usually through openings in manhole lids or improper connections that convey stormwater directly into the wastewater collection system. Rainfall derived infiltration is the infiltration component that is directly correlated with rainfall events.

Flow Discretization

Now let's discuss the method for determining these flow components. The following analysis allows for the separation of flow into its discrete categories.

Dry-Weather Analysis

The dry weather analysis allows for the determination of the BWWF and GWI components. This is where the rainfall data starts to come into play. The BWWF and GWI components do not have a direct correlation to rainfall so it necessary to find dry days in your flow monitoring analysis. You may develop some parameters for determining what constitutes a *dry* day (i.e. no rainfall within 48 hours) but the best way to determine this is to review the flow data and identify periods where the flow reverts to a standard average. Statistical analyses can be done to determine these periods mathematically, but that is beyond the purview of this text. **Figure 7-2** should help with this explanation. Note the peaks that correlate with the rainfall events and how they gradually subside, sometimes over several days.

Once you isolate the dry weather flow, you can separate the GWI from the BWWF. This is done using various methods, but one common method is called the Stevens-Schutzbach Method (SSM). This method calculates the GWI using the following equation:

$$GWI = \frac{0.4(MDF)}{1 - 0.6\left(\frac{MDF}{ADF}\right)^{ADF^{0.7}}}$$

- *MDF = Minimum Daily Flow*
- *ADF = Average Daily Flow*

The minimum daily flow (MDF) usually occurs around 3 am, when most people are asleep and there is minimal contribution from residences. High MDF suggests that there is significant influence from industrial or commercial sources or that the GWI is high. The average daily flow (ADF) is the average flow rate for the analysis period. These concepts are detailed in **Figure 7-2**.

Wet-Weather Analysis

The RDI/I contribution is determined by reviewing the flow response to rainfall events. **Figure 7-2** shows this flow component graphically. It is defined as the flow that is added to the system during a rainfall event, so it can be determined by subtracting the BWWF and GWI from the total flow. It is difficult to accurately separate the rainfall derived inflow from the rainfall derived infiltration, so this is usually not done during an analysis.

Flow Monitoring Analysis Results

Below is a description of some common information that is gathered during the flow monitoring analysis.

GWI/idm

GWI per inch diameter mile (GWI/idm) is a standard parameter used by the EPA to categorize a sub-basin. Excessive values for this parameter are usually 1,500 GWI/idm or higher (Metcalf & Eddy, Inc., 1981).

R-Value

The R-value represents the amount of rainfall that enters the wastewater collection system during a storm event. This is calculated using the following equation:

$$R - Value = \frac{RDII\ Volume}{Rainfall\ Volume}$$

It is usually represented as a percentage and can be used to approximate the volume of stormwater that will enter the system during any given storm event.

Peak Wet Weather Flow Factor

This peak factor is the result of the inflow, or immediate response to the rainfall event. You can see the peak identified in **Figure 7-2**.

$$Peak\ Factor = \frac{Maximum\ Flow\ during\ Storm\ Event}{Average\ Flow\ during\ Dry\ Period}$$

RDI/I per Linear Foot

The RDI/I per linear foot is useful in identifying potential areas for rehabilitation. If you have an area with a very high RDII/LF calculation, then it will be less costly to rehabilitate this area that it may be for a region that is producing a higher level of RDI/I but has a lot more sewer mains that would need to be inspected.

Figure 7-2: Wet Weather Flow Parameters

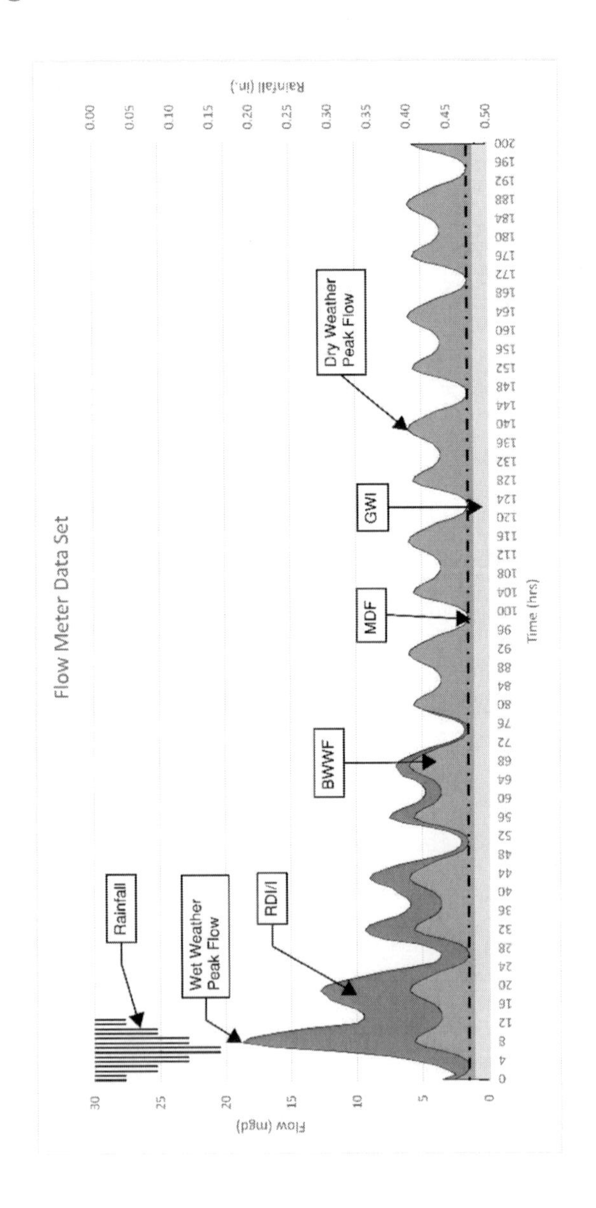

SSES

Sanitary Sewer Evaluation Surveys (SSES) consist of detailed inspections of the wastewater collection system to identify areas that require or would benefit from rehabilitation. The flow monitoring analysis assists in identifying these areas. Manholes and pipes are inspected using closed-circuit television (CCTV) and categorized based on the severity and quantity of defects. The North American of Sewer Service Companies (NASSCO) Pipeline Assessment Certification Program (PACP) and Manhole Assessment Certification Program (MACP) standards are typically used for this categorization. This is very useful in mitigating high RDI/I contributions to the system.

Sewer Model Development and Calibration

I briefly discussed the physical parameters necessary to create the sewer model. Now I want to provide an overview of the purpose of this model and some considerations in finalizing its development. This will not be program specific, so I will not go into much detail regarding input data. I should also point out that sewer modeling could be the subject of an entire book that I'm not qualified to write. This is a specialty that some spend an entire career focused on. However, I can provide some general information based on my limited experience with modeling that should at least give you a general understanding of the concepts.

Base Loading

This will be the BWWF and GWI that we discussed previously. You will typically determine a starting point for your base loading by reviewing customer data, population data, and some combination. You will also need to model the groundwater conditions to ensure they are included in the flow. Diurnal patterns can be set for each loading point (usually a manhole) that are determined based on a review of the flow meter data.

Dry Weather Calibration

Calibration basically means adjusting some parameters (i.e. pipe roughness, base loading) to match the flows in the model at a metered location to the flows that the actual physical meter picked up. By doing this, you are adjusting the model to mimic actual field conditions.

Wet Weather Calibration

Wet weather calibration is much more complex. The rainfall response is usually modeled by using the RTK method. This consists of developing parameters that are used to calculate a unit hydrograph to represent the system response to a rainfall event. Following is a brief discussion of the RTK method.

RTK Unit Hydrograph Method

- ***R**: The Fraction of Rainfall that enters the Sewer System*
- ***T**: The Time to Peak*
- ***K**: The Ratio of Recession Time to time to Peak*

Three-unit hydrographs are created based on three sets of RTK parameters. The first unit hydrograph represents the stormwater runoff that immediately enters the sewer system. This is usually the inflow. The second unit hydrograph represents the flow from both inflow and infiltration. The third unit hydrograph represents the long-term response and is sourced by infiltration. These three-unit hydrographs are then added to create a composite hydrograph to represent the system response to RDI/I. Flow meter data is used to determine the RTK parameters that provide the most accurate representation of the impact of real-world storm events on the sewer system.

Capital Improvement Projects

Now that you have a computer model setup that provides a good representation of the actual wastewater collection system, you can start to look at areas that may need improvements today, or within a specified planning horizon (i.e. 20 years). You will use this data to recommend projects to the system owner and let them know when they should implement those projects.

The analysis of the existing system is relatively straightforward. Once you have some design parameters (i.e. pipe must not be more than 75% full at any given time), you can identify areas where these parameters are not met, and possible upgrades are needed. Then you can start looking into the future. Usually this is done by developing flow projections based on planned developments and expected population growths. Keep in mind that these projects usually cost millions of dollars to implement, so it is important that you are identifying the right ones. You can imagine the problems that would be caused by identifying a multi-million-dollar project and later finding that it was inadequate to meet the capacity or unnecessary. That's why modeling is a specialty and you should defer to experienced personnel for assistance. The previous discussion provides an overview, but by no means prepares you for a high-level sewer analysis.

Design Elements (Gravity)

Pipes

Pipe Material

The most commonly used pipe materials can be found in **Table 7-2**. PVC is the most cost-effective option but is not always feasible depending on the depth and size of the pipe. **Table 7-2** provides typical materials, size ranges and special case information, but pipe materials can vary with region and municipality, so it is very important that you thoroughly review the regulations for the utility provider.

Table 7-2: Wastewater System Pipe Materials

Material	Size Range	Comments
Polyvinyl Chloride (PVC)	8” to 15”	Use when sufficient cover is available
Ductile Iron Pipe (DIP)	>8”	Insufficient cover or separation
Reinforced Concrete Pipe (RCP)	>12”	Large mains

Pipe Sizing and Minimum Slope

Pipe sizing is done using Manning’s equation for basic project design. The level of detail used for sewer system design will vary depending on the size of the project and requirements of the municipality. In some cases, basic spreadsheet calculations will be used. In others, a full model will be developed. The minimum pipe size is 8 inches for a sewer main, and this typically has more than sufficient capacity for commercial and residential developments. However, when you are designing or analyzing larger systems, it is important to develop a model. Flow in a large system can be complex, and spreadsheet calculations will not suffice unless you are unreasonably conservative. Following are examples of minimum slope and capacity calculations.

Minimum Slope

The flow must maintain a minimum velocity of 2 feet per second to prevent deposits from settling on the bottom. This is called the self-cleaning velocity, and it is calculated using Manning's equation with a typical roughness factor being 0.013. **Table 7-3** shows the minimum slopes recommended by the *Ten State Standards*. Following is a sample calculation for a 10-inch pipe. Note that the recommended values provided by the *Ten State Standards* are a little more conservative since this calculation assumes full pipe flow:

$$V = \frac{1.49}{n} R^{\frac{2}{3}} S^{\frac{1}{2}} \rightarrow S = \left(\frac{V}{\frac{1.49 R^{\frac{2}{3}}}{n}} \right)^2$$

$$R = \frac{A}{P} = \frac{\pi r^2}{\pi d} = \frac{\pi * \frac{5}{12}^2}{\pi * \frac{10}{12}} = 0.351$$

$$S = \left(\frac{V}{\frac{1.49 R^{\frac{2}{3}}}{n}} \right)^2 = \left(\frac{2}{\frac{1.49(0.351)^{\frac{2}{3}}}{0.013}} \right)^2 = \mathbf{0.25\%}$$

Pipe Capacity

The pipe capacity can also be calculated using Manning's equation. Following is a sample calculation of the capacity for a 10-

inch pipe at a 1% slope. Note that the slope can have a significant impact on the capacity.

$$V = \frac{1.49}{n} R^{\frac{2}{3}} S^{\frac{1}{2}} = \frac{1.49}{0.013} * 0.351 * \left(0.01^{\frac{1}{2}}\right) = 4.03\ ft/s$$

$$Q = VA = 4.03 * 0.55 = 2.20\ cfs = \mathbf{986.03\ gpm}$$

Table 7-3: Minimum Slope[14]

Diameter (inches)	Minimum Slope (%)
8	.40
10	.28
12	.22
15	.15
16	.14
18	.12
21	.10
24	.08
27	.067
30	.058
33	.052
36	.046
39	.041
42	.037

Note that these are absolute minimum slopes and assume a pipe full flow condition. Often your firm will have a general standard minimum slope such as 0.005 ft/ft. The reason for this is that though lower slopes are possible in special conditions, caution should be taken to ensure that the slope is constructible within reasonable tolerances. It should also be noted that the steeper the slope, the higher the capacity, so using the minimum slope should be avoided if possible.

[14] (Greater Lakes - Upper Mississipi RIver Board of State and Provincial Public Health and Environmental Managers, 2014)

Minimum Cover

The standard minimum cover requirement for a sanitary sewer PVC pipe is 3 feet. This may vary depending on the loading conditions above the pipe. Any pipe with less than 3 feet of cover should be DIP or RCP. Cover should never be less than 1 foot. During the design process, it is important to give yourself a little bit more than 3 feet of cover to allow for variations in the grade over the pipe and construction tolerance. If you design a pipe with exactly 3 feet of cover, it is possible that the installed pipe will have less than 3 feet of cover in some areas and you will not receive the necessary permit to operate until this is corrected.

Minimum Separation

The standard minimum vertical separation is 18 inches. It is preferred that the wastewater pipe is installed below the water main to prevent contamination. It is also necessary to provide a minimum of 10 feet of horizontal separation between the sewer main and any proposed or existing water main. In cases where the 18 inches of vertical separation cannot be met, you should coordinate with the regulatory agency to determine methods of pipe encasement that will allow a reduction of this requirement. Caution should also be taken when you have large storm drainage pipes above your sanitary sewer. In these cases, you may need to increase the separation or use DIP to prevent damage from the vertical load.

Easements

Easements will be required in any case where the public entity will maintain ownership and maintenance responsibilities of the wastewater main. A typical easement width is fifteen feet, but this will vary depending on agency guidelines, depth, and size. The main will need to be accessible in any case where maintenance is needed, and an adequate easement is needed to allow for this maintenance.

Anti-Seep Collar

Anti-seep collars are used whenever there is a potential for the seepage of water to undermine the interface between the pipe and the soil. The collars are designed to increase the distance that the water must travel along the pipe. There is not a lot of consensus regarding the efficacy of this method, but it is still commonly used in design. These are often used in areas where there is a high groundwater table or saturated soils such as wetlands or stream crossings.

Manholes

A manhole is used at any change in direction in a sanitary sewer gravity system. Manholes are typically 4 feet in diameter (outside), but the size may be larger depending on the pipe diameter. Manholes are used to access the sewer system for inspection and cleaning.

Manhole Spacing

Manhole spacing varies depending on pipe size, but a good rule of thumb is to space manholes at a maximum of 400 feet. Manholes are access points for inspection and cleaning, so excessive distances between them makes these tasks exceedingly difficult.

Manhole Types

In cases where the invert into the manhole is greater than 24 inches above the bottom of the manhole, an outside or inside drop will be necessary. This is called a drop manhole, and it will facilitate the flow through the manhole without splashing. **Figure 7-3** shows a typical manhole and drop manhole.

Figure 7-3: Typical Manholes

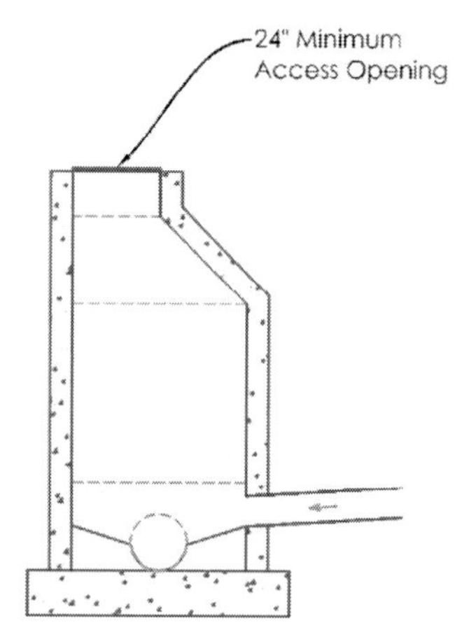

Typical Manhole

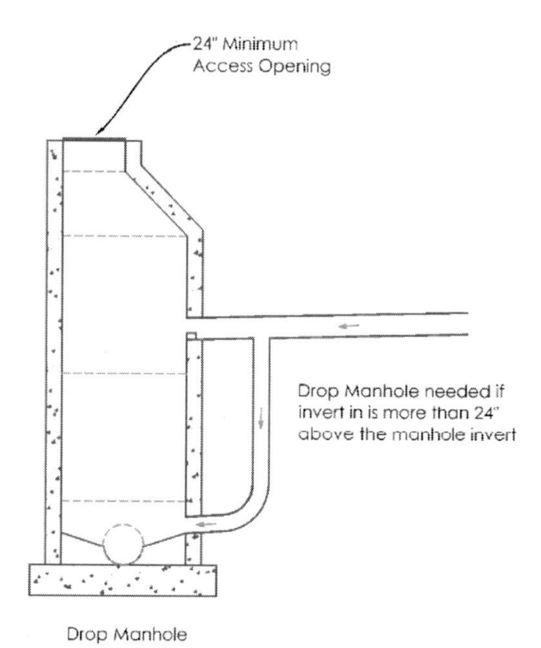

Drop Manhole

Design Elements (Pressure)

Pumping Systems

For the conveyance of wastewater, it is preferable to transport flow via gravity systems. However, you will often come across circumstances where this is not feasible. Not only will flow need to overcome the elevation difference, but it must also overcome energy losses in the system.[15] Following is a basic overview of pumping system terminology and concepts.

Head

Static Head

The static head is the sum of elevation and/or pressure differences between the initial water level and the control point. The control point is usually the discharge location, but it may also be a high point along the force main.

In wastewater pump station systems, flow will typically be discharged to a gravity manhole, so there will be no pressure difference, but in some cases, such as when you will tie directly to another force main, you will also need to consider the pressure difference that must be overcome by the pump. An illustration of this concept can be found in **Figure 7-4**.

Head Losses

In addition to overcoming elevation and pressure differences, the pump must also be designed to overcome major and minor head losses. Depending on the distance and pipe size/material, this can be

[15] While pumping systems are used in water and stormwater systems, they are less prevalent, and their design is usually more specialized. For this reason, I only go in-depth regarding wastewater pumping systems.

a significant component of the pump sizing calculations. These losses will include both friction losses and minor losses from bends, valves, and other appurtenances. **Figure 7-4** shows this concept.

Total Dynamic Head

The total dynamic head (TDH) is a combination of the static head and head losses in the system. This is the total head that the pump must be designed to overcome in order to move water from its initial water level to its final elevation (or a high point in the system). **Figure 7-4** represents a basic pump station and force main diagram. Notice that the decline in HGL is equal to the head loss, and the TDH is the sum of the static head and head loss.

Net Positive Suction Head

While the total dynamic head is the head that the pump needs to overcome, the Net Positive Suction Head (NPSH) is the minimum head that must be maintained to prevent pump cavitation.

Cavitation

Cavitation can cause significant damage to pumps over time. It results from significant pressure differences that cause rapid changes in the state of the wastewater, from liquid to vapor and back to liquid. The vaporization creates bubbles (cavities) that implode and release high levels of energy to the pump casing and impeller. There are two types of cavitation: suction and discharge.

Figure 7-4: Total Dynamic Head

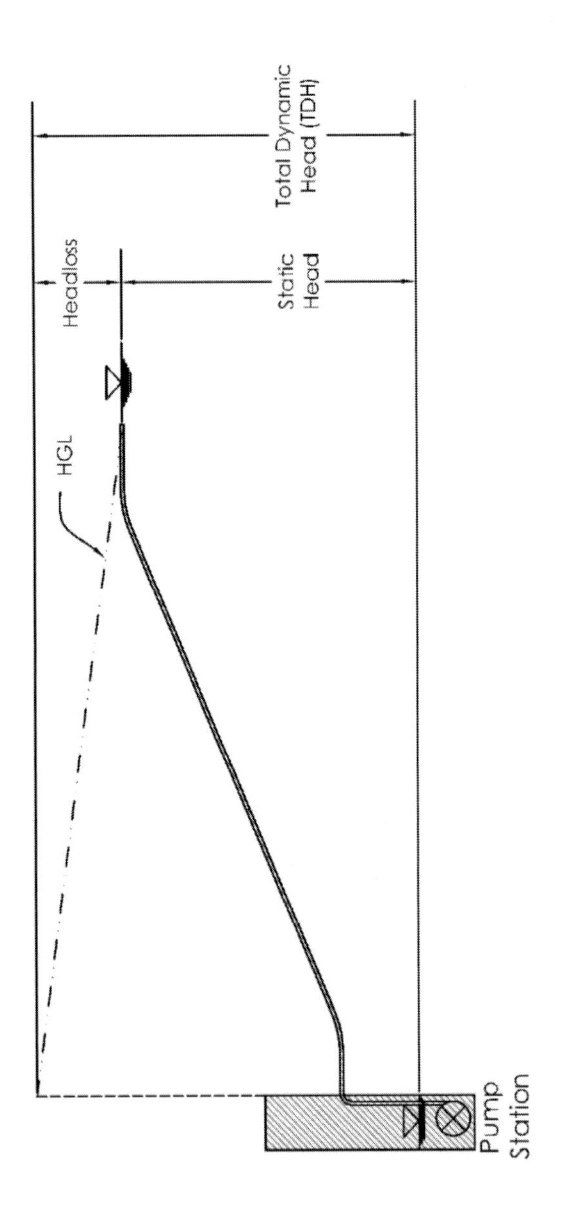

Suction cavitation results from pressures that are too low, which means the NPSH condition discussed previously was not met. This is not usually a concern for submersible pumps (which are often used in smaller stations) because they do not usually have a suction component and they are constantly submerged under a head of water.

Discharge cavitation results from pressures that are too high. This means the pump is having to work too hard and is unable to push all the water out. Both of these conditions can be avoided with proper pump selection which will be discussed shortly.

System Curve

The system curve is a graphical representation of the system components for the conveyance of the fluid and does not have anything to do with pump characteristics. This curve is developed by plotting the Total Dynamic Head at various flow rates. The shape of the curve results from the fact that as the flow rate increases, minor and major losses will increase as well. **Figure 7-5** shows a graph of a typical system curve. The static head is the head at a flow rate of zero. The curve indicates, for example, that if the flow in the system is moving at a rate of 600 gallons per minute (gpm), the total dynamic head (the head that must be overcome by a pump that is producing this flow rate) is approximately 100 feet. This curve will be used in conjunction with the pump curve for pump selection.

System Curve Example Problem

Let's work through a very basic example of how to develop the system curve. This will also familiarize you with the major and minor loss equations. We are designing a pump station that must move wastewater from a starting elevation of 100' in the wet well, discharging to a gravity manhole at an elevation of 150' (this is the highest elevation of the force main). The water will be transported through a 6-inch DIP force main that consists of one flanged 90-degree bend, two threaded regular 45-degree bends, one gate valve, one check valve and one ball valve (all valves are open).[16] The total distance will be 2,000 feet.

To develop the system curve, we need to determine the Total Dynamic Head for various flow rates. First, we can easily calculate the static head by subtracting the starting elevation from the discharge elevation:

$$150' - 100' = 50' \; static \; head$$

The major and minor losses will vary with flowrate and can be calculated using the equations discussed previously. The C value is 140 for DIP. The sum of the minor loss coefficients (k) found in **Table 5** will be 1.3. This calculation is much easier to do using a spreadsheet. We'll start by determining our parameters:

- *Static Head = 50 feet*
- *Diameter = 6 inches*
- *Radius = 3 inches*
- *Length = 2,000 feet*
- *k = 1.3*
- *C = 140*

---

[16] This is not meant to be an accurate representation of the valves you will see in a typical pump station design. This is only meant to assist with an understanding of minor loss calculation basics. Also note that there are other methods, such as the equivalent length method, used for minor loss calculations.

Then we will build a table with the calculations:

Table 7-4: System Curve Example Calculations

Flow Rate	Area	Velocity	h_m	h_f	TDH
gpm	sf	ft/s	ft	ft	ft
0	0.196	0.00	0.000	0.000	50.0
50	0.196	0.57	0.006	0.507	50.5
100	0.196	1.13	0.026	1.829	51.9
150	0.196	1.70	0.058	3.873	53.9
200	0.196	2.27	0.104	6.595	56.7
250	0.196	2.84	0.162	9.965	60.1
300	0.196	3.40	0.233	13.962	64.2
350	0.196	3.97	0.317	18.570	68.9
400	0.196	4.54	0.415	23.774	74.2
450	0.196	5.11	0.525	29.562	80.1
500	0.196	5.67	0.648	35.924	86.6
550	0.196	6.24	0.784	42.851	93.6
600	0.196	6.81	0.933	50.334	101.3
650	0.196	7.38	1.095	58.368	109.5
700	0.196	7.94	1.270	66.945	118.2
750	0.196	8.51	1.457	76.059	127.5

Now we can plot the system curve with the flow rate along the x-axis and the total dynamic head along the y-axis:

Figure 7-5: System Curve

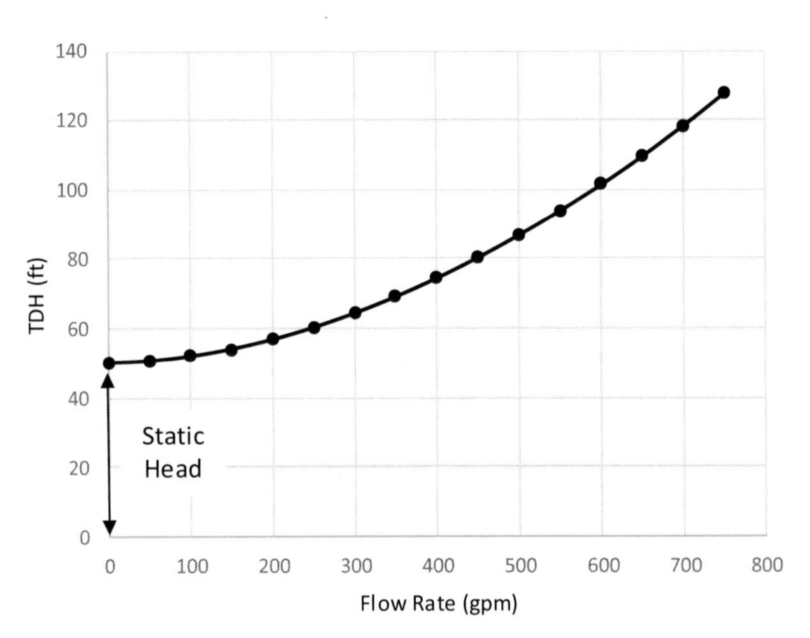

Note that as the flow rate increases, the total dynamic head will increase also. This is because as the flow rate increases, the velocity will increase which results in greater major and minor losses. This is a very important concept to understand and retain.

Pump Performance Curve

The pump performance curve provides information on how much head the pump can overcome at various flow rates. This can be plotted on the same graph as the system curve to determine the operating point. To understand the concept of the pump curve, imagine you are riding a bike up a hill. As the hill gets steeper, putting forth the same effort will result in lower speeds. This is the same with a pump. As the head that the pump needs to overcome increases, the pump pushes the wastewater at slower velocities. The flow rate that the pump operates at will also impact the selection of the force main which will be discussed later. **Figure 7-6** shows an example of a pump performance curve.

Efficiency Curves

Efficiency curves indicate the efficiency that the pump is operating at for a given flow rate. You will want the pump to operate at the highest efficiency possible.

NPSH Curve

The NPSH curve indicates the minimum head that must be maintained in the system to prevent damage to the pump due to cavitation.

Pump Selection

Pump selection can be completed by combining the system and efficiency curve and the pump performance curve to determine if the operating point is ideal for the proposed pump. **Figure 7-6** provides an example of what a typical pump curve selection graph will look like. The first step in pump selection is determining a desired operating point. This will be based on the peak inflow rate to the pump station. The pump must produce a flow rate that is greater than the peak inflow to prevent the wet well from overflowing. The pump manufacturer will usually assist with pump selection once you provide the system curve and a desired operating rate. It is important to select a pump that operates with reasonable efficiency. I will go into more detail on pump selection in the design process section.

Figure 7-6: System Curve and Pump Curve Interaction

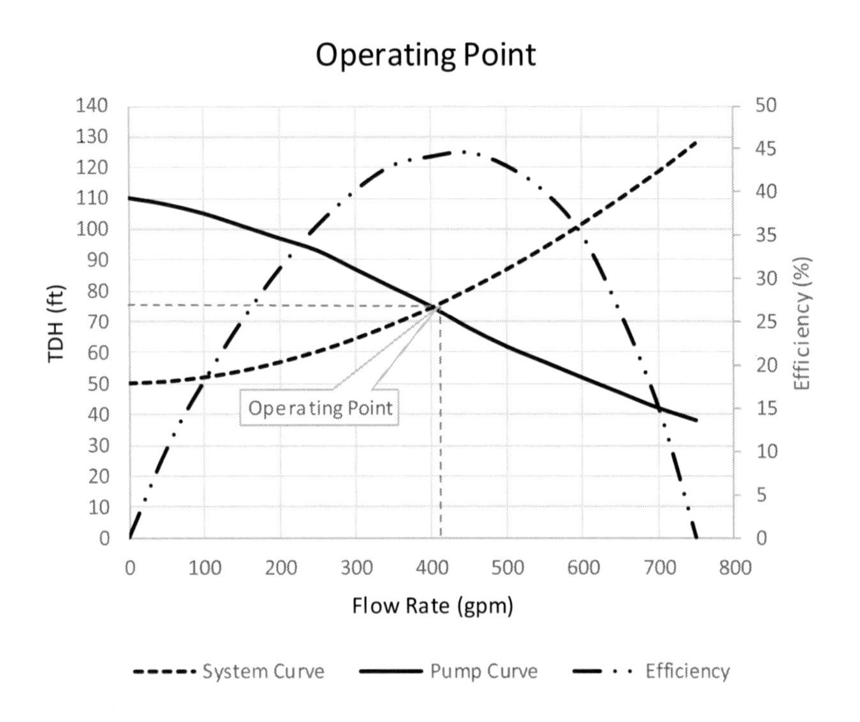

Operating Point

The point where the system curve crosses the pump curve is called the operating point. This is the flowrate that the pump will operate at for that system.

Pump Station

The pump station, also called a lift station, is designed to overcome the total dynamic head (head losses and elevation change) in a wastewater conveyance system. There are several detailed components in a wastewater pump station, but I will focus on the primary components that help to convey the overall concept: these being the wet well, pumps, and force main.

Wet Well

Wastewater will be conveyed to the pump station via gravity sewer, or in some cases, via a force main from another pump station. This wastewater will flow into the wet well. In most cases (for smaller pump stations), the wet well will house the pumps, though there are some larger pump stations that house the pumps in a dry well for ease of maintenance. In either case, the wastewater discharges to the wet well and is pumped out through the force main, where it will eventually discharge to a gravity manhole or another pump station. See **Figure 7-7** for a basic illustration of this concept.

Pumps

Each pump station must contain a minimum of two pumps. This is for redundancy - the station must be able to maintain adequate capacity with the largest pump out of service (this is called Firm Capacity). There are different types of pumps used depending on whether they will be housed in the wet well or dry well. In most cases, you will rely on the pump manufacturer to assist with pump selection, so I will not go into detail regarding pump types, but it is important to understand the basic characteristics of pump selection based on the desired operating point. This pump selection concept was previously discussed but will be covered in more detail in the design process section.

Operational Settings

Operational settings control when the pumps operate and when alarms sound based on the elevation of the wastewater in the wet well.

Alarms for pump stations without full-time maintenance crews are remotely monitored. **Table 7-5** contains an overview of the operational settings for a typical small pump station. Note that in larger pump stations, the pump on/off elevations become more complex, but the concept is the same.

Table 7-5: Operational Settings

Low Level Alarm	This alarm notifies personnel when the wet well level falls too low.
Pumps Off	All pumps are turned off when this level is reached.
Lead Pump On	The Lead pump turns on when this level is reached.
Lag Pump On	The Lag pump is turned on when this level is reached. Note that if both pumps must run at the same time, the pump station has exceeded its capacity.
High Level Alarm	This alarm notified personnel when the wet well level falls too low.

Figure 7-7: Typical Pump Station

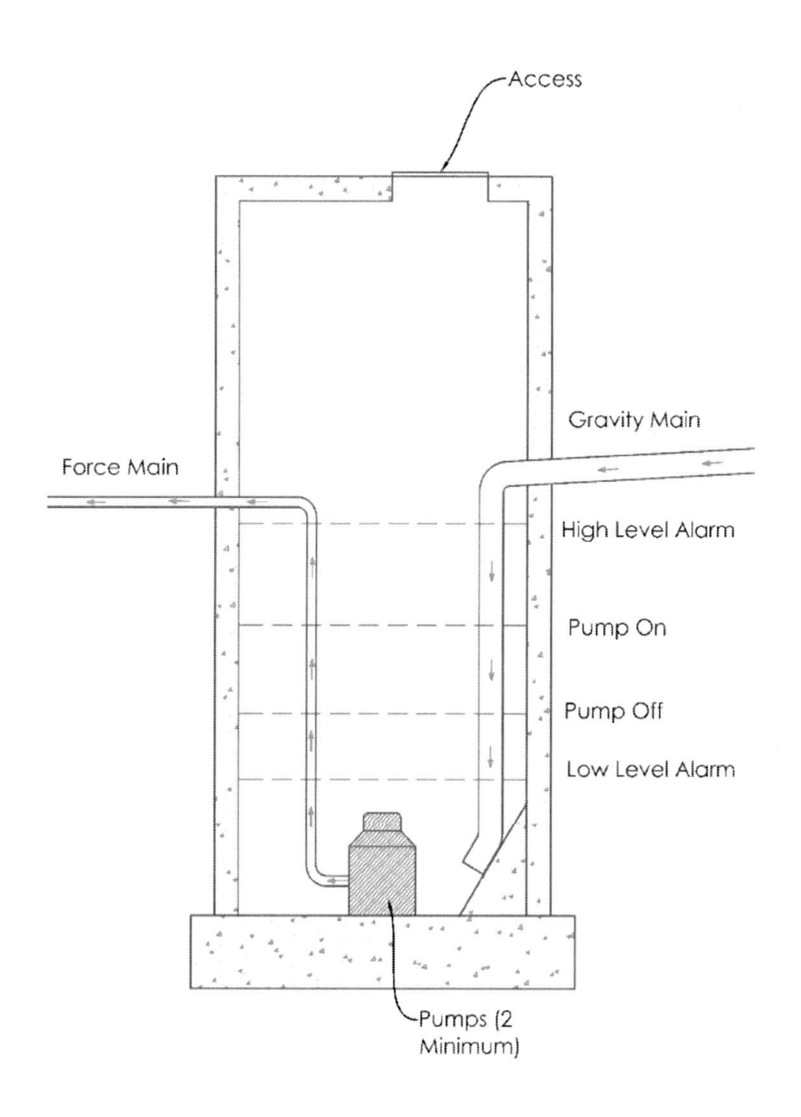

Force Main

The pipe that carries wastewater from a pump station to its discharge location (usually a manhole or another pump station) is called a force main. This will usually be made of DIP or PVC. The force main design will use similar concepts as the water main design that we discussed previously. The selection of the force main size will depend on the desired operating condition of the pump. It is preferable to maintain velocities between 2 feet per second and 5 feet per second. Flow velocities below 2 feet per second will result in deposits on the invert of the force main and flow velocities above 5 feet per second can indicate inefficient design due to the high friction losses. Remember that the operating condition of the pump is dependent on the force main size, so there will be some iteration in this design. To check the force main velocities, you can use the Hazen-Williams equation.

Figure 7-8: Typical Force Main Profile

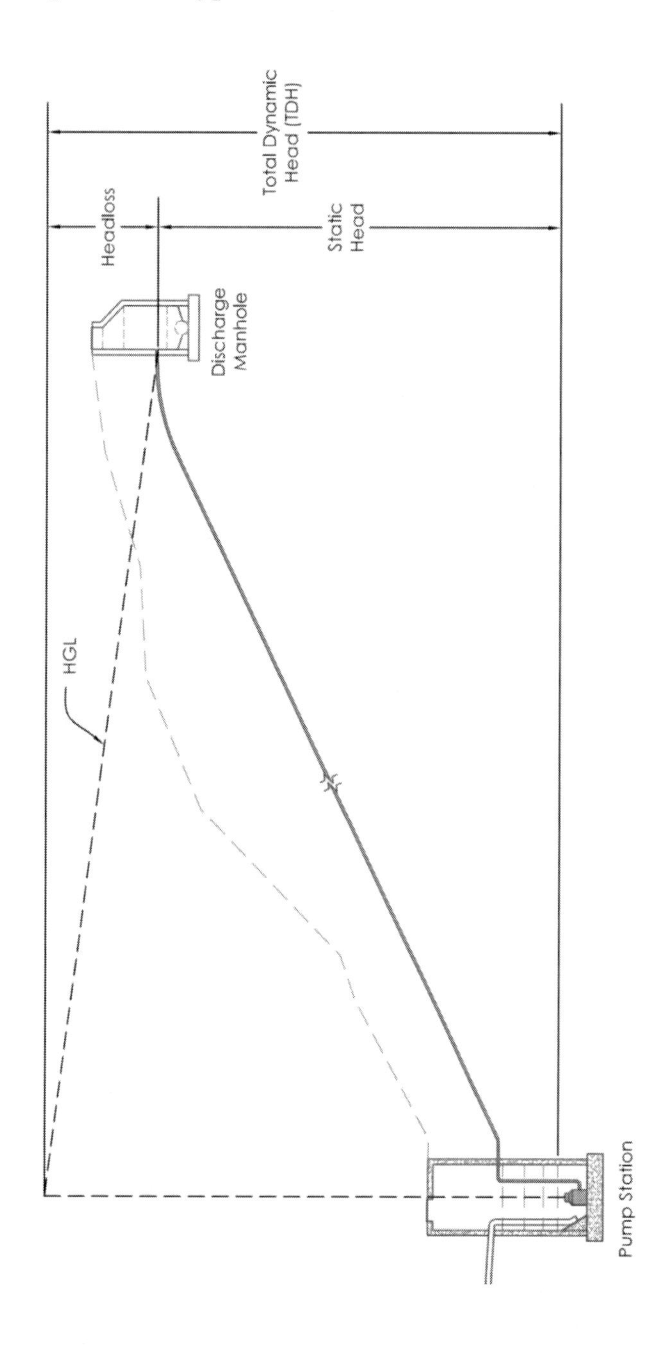

Figure 7-9: Control Point

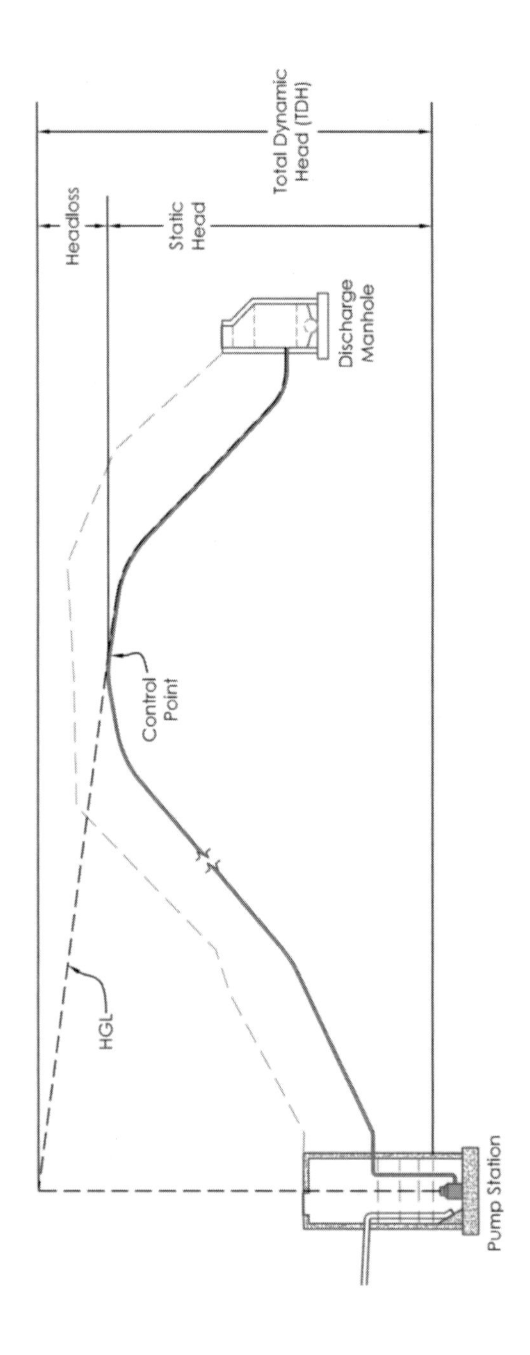

Design Process (Gravity)

The first step in the sanitary sewer system design process is to determine where you will discharge the new sanitary flows. You will typically either discharge to an existing manhole or pump station. Once you've identified the discharge location, you must determine the minimum allowable invert for your design. Then you must determine the areas that must be served by the proposed sewer system. These will be the buildings (residential, commercial, or industrial) that will produce sanitary flow to the system. This process is not too complicated for smaller projects, but it can be an important factor in your design. Identify your potential sewer route early and determine if gravity will be possible. Also remember the chapter on grading. You may need to consider the sewer design when determining the proposed grading. Wastewater mains plan and profile views showing some general design concepts can be found at the end of the chapter.

Design Process (Pressure)

If you've analyzed the site and determined that a pump station will be necessary, you will first need to locate it. This will usually be done in conjunction with the municipality since they will often take ownership after the pump station is constructed. The pump station will generally be located at the lower portion of the site, so that the proposed structures can connect to it via a gravity system.

We've already discussed some basic information regarding pump stations, so this section will focus on some more specific design concepts. To understand the basic design process for pressure systems, let's walk through an example for a minor[17] pump station design. Your

[17] Large pump stations have several additional complexities that go beyond the scope of what I'm trying to convey in this book.

firm will probably have a spreadsheet set up to guide you through this, but I encourage you to create one yourself. That way you don't blindly input data without fully understanding what it means. Also, understanding this basic design brings a lot of concepts together that we've discussed previously. Don't focus too much on the specifics, just make sure you understand the general concept.

We'll design a pump station to serve a 600-lot residential development (single-family homes). The ground elevation where the pump station will be installed is 230 ft. The force main will be 2,000 feet of ductile iron pipe with the following key elevations at the existing ground along the alignment:

- *Elev: 250 ft @ Station 02+00*
- *Elev: 300 ft @ Station 12+00*
- *Elev: 350 ft @ Station 17+50*
- *Elev: 275 ft @ Discharge*

The invert into the pump station will be set at 220 feet. This pipe connects upstream to the manhole with the lowest rim elevation in the system. The pump station will have two submersible pumps. The minimum depth of wastewater in the wet well should be 4 feet.

Now let's look at the equation that is the basis for the overall design concept:

$$\frac{\Delta V}{\Delta T} = Q_{out} - Q_{in}$$

- ΔV = *Change in Volume*
- ΔT = *Change in Time*
- Q_{out} = *Flow Out (Pumped)*
- Q_{in} = *Flow In (Gravity)*

First let's calculate the Flow In. This is the flow produced by the 600-lot development. From **Appendix D,** we can see that the typical average daily loading rate for a single-family home is 300 gallons per day (gpd). So the total average daily flow (ADF) into the pump station will be 180,000 gpd or 125 gpm. This is the average daily flow. Don't forget the diurnal patterns we discussed earlier. This will not be a steady flow of 125 gpm, though some of our design will make that assumption. You will need to make sure that you have adequate capacity for the peak flow condition. A peak factor of 2.5 is the most common. So, we have the following parameters for Flow In (Q_{in}):

- $Q_{in(ADF)} =$ *125 gpm*
- $Q_{in(Peak\ Flow)} =$ *312.5 gpm*

Let's talk a little more about how the pump station operates, because it is important to understanding the next design steps. The pump station will be designed with two pumps, but you don't want to design the pump station to need both pumps at the same time. The pump station should be able to function with only one pump and the second pump provides redundancy in case the other pump fails. The pumps will usually be designed to alternate to decrease the run times. There are multiple elevations that must be set in the wet well as shown in **Figure 7-7** and listed below.

- *Top of Wet Well*
- *Invert of Wet Well*
- *Pumps Off*
- *Lead Pump On*
- *Lag Pump On*
- *High Level Alarm*

We can start by setting some of these elevations based on the data we were provided. We know that the ground elevation is 230 ft so this will be the top of the wet well. We will set the High-Level Alarm 2 feet below the invert of the gravity main; 218 ft. To determine the wet well invert and pump operating elevations, we will need to calculate the necessary volume. To do this, let's think about how the pump station operates. Flow will enter from the gravity main into the wet well. The elevation inside the wet well will increase until it reaches the Lead Pump On elevation. The pump should lower the wet well elevation until it reaches the Pump Off elevation, when the cycle will begin again.

So, the next step will be to determine a cycle volume. We do this by deciding how many cycles we want per hour; a standard is somewhere between 2 and 8 cycles. For this example, let's shoot for 4 cycles per hour. This means that one cycle must complete in 15 minutes. We've already determined that the ADF is 125 gpm and the Peak Flow is 312.5 gpm. Since our peak flow is 312.5 gpm, our pump must be able to exceed that flow rate in order to pump down the wet well. Let's select a target pumping rate of 325 gpm. Now we can look back at the basic equation:

$$\frac{\Delta V}{\Delta T} = Q_{out} - Q_{in}$$

We will use the ADF for the flow in for the purpose of the cycle volume calculation. Just keep in mind that the actual flow in will fluctuate according to the diurnal pattern. So, while we are calculating for a specified number of cycles per hour, this will not be a fixed pattern, it will change depending on the time of day. Here is the calculation for cycle volume:

$$Cycle\ Time\ (T) = TIme\ to\ Fill + Time\ to\ Pump\ Down$$

$$Time\ to\ Fill = \frac{V}{Q_{in}}$$

$$Time\ to\ Pump\ Down = \frac{V}{Q_{out} - Q_{in}}$$

$$T = \frac{V}{Q_{in}} + \frac{V}{Q_{out} - Q_{in}} \rightarrow V = \frac{T}{\left(\frac{1}{Q_{in}} + \frac{1}{Q_{out} - Q_{in}}\right)}$$

$$V = \frac{15}{\left(\frac{1}{125} + \frac{1}{325 - 125}\right)} = 1154\ gallons$$

Let's convert this to cubic feet to simplify the next calculations:

$$1154\ gal \left(\frac{1\ ft^3}{7.48\ gal}\right) = 154\ ft^3.$$

Since we are assuming an 8' diameter wet well, we know that:

$$Area = \pi r^2 = \pi(4^2) = 50.27\ ft^2$$

That means we would need 3 feet between the pumps off and pump on elevations:

$$V = A * H = 50.27\ ft^2 * 3.1\ ft = 150\ ft^3$$

It's typical to have the Lag Pump On elevation 1 foot above the Lead Pump On elevation and the High Alarm elevation 1 foot above that. Since we know the High Alarm elevation is set at 218 feet, we can determine that the Lag Pump On elevation should be 217 feet and Lead Pump On elevation should be 216 feet. We need 3.1 feet in the cycle volume, so the Pumps Off elevation will be 212.9 ft. We've specified that there should be a minimum of 4 feet between the Pumps Off elevation and the wet well invert, so the invert will be 208.9 ft. This should be all the parameters needed for the basic pump station design.

This example should help you understand the concept of pump station design. It's more important to understand the general concepts than any specific rules of thumb. Your firm will likely have some of its own general rules that will help guide you. Just remember that the pump station must have the capacity to pump the peak inflow to the station. That means your operating point must be higher than the peak hourly flow into the pump station.

Figure 7-10: Wastewater Gravity Main Plan (Commercial)

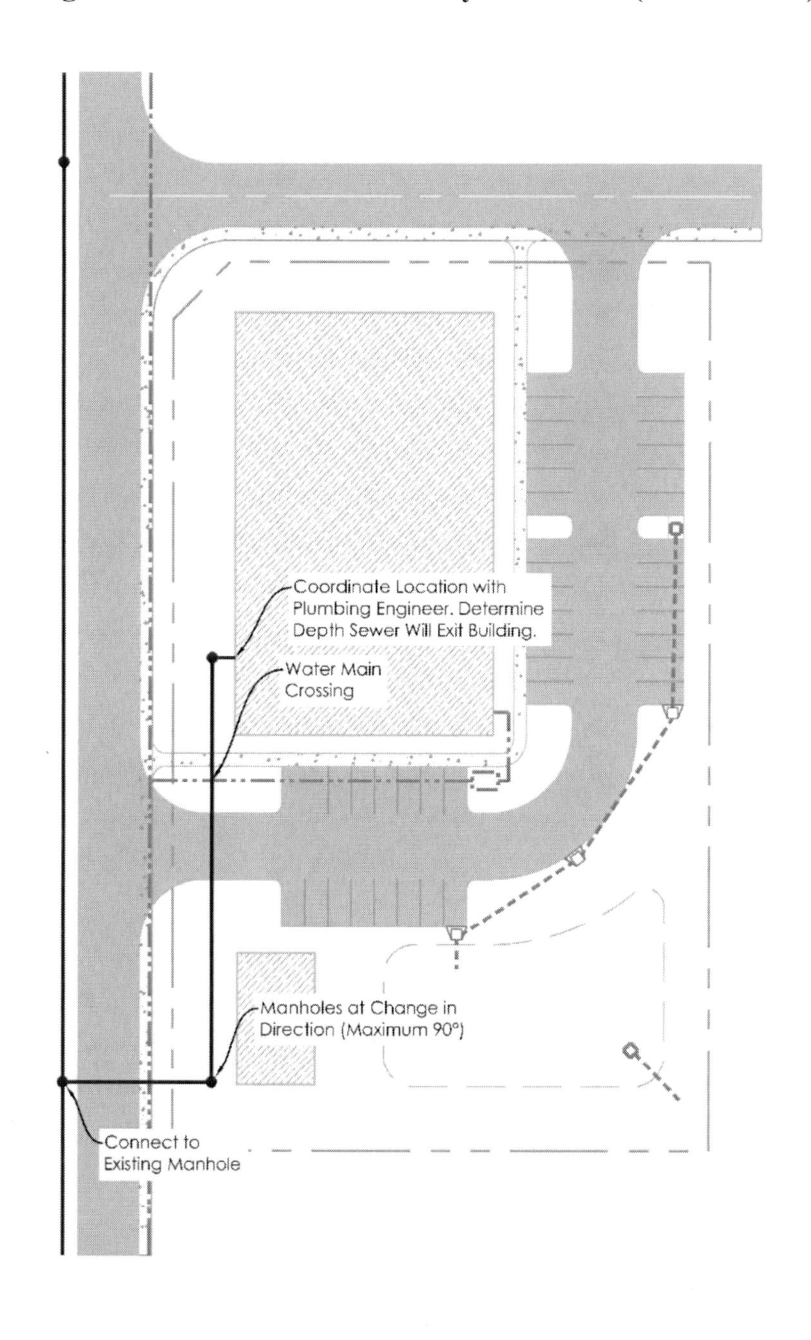

Figure 7-11: Wastewater Gravity Main Plan (Residential)

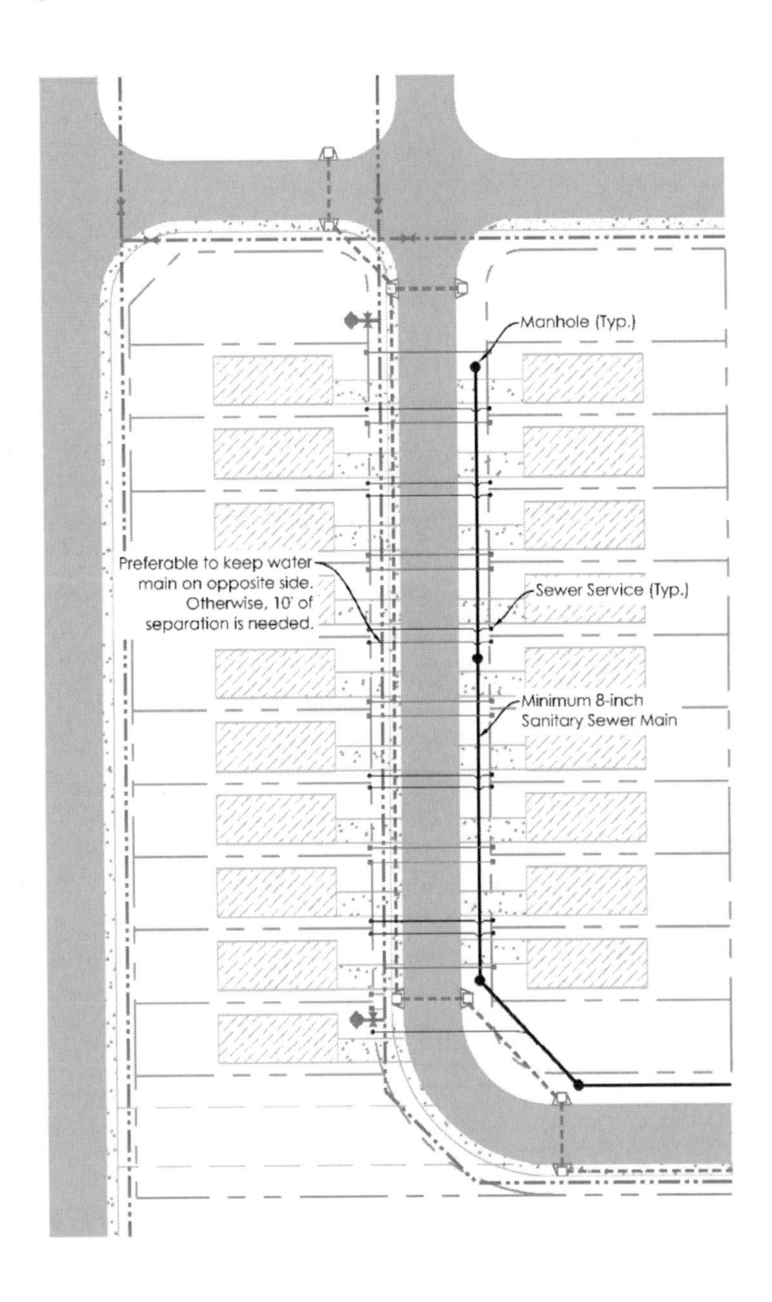

Figure 7-12: Wastewater Gravity Main Profile

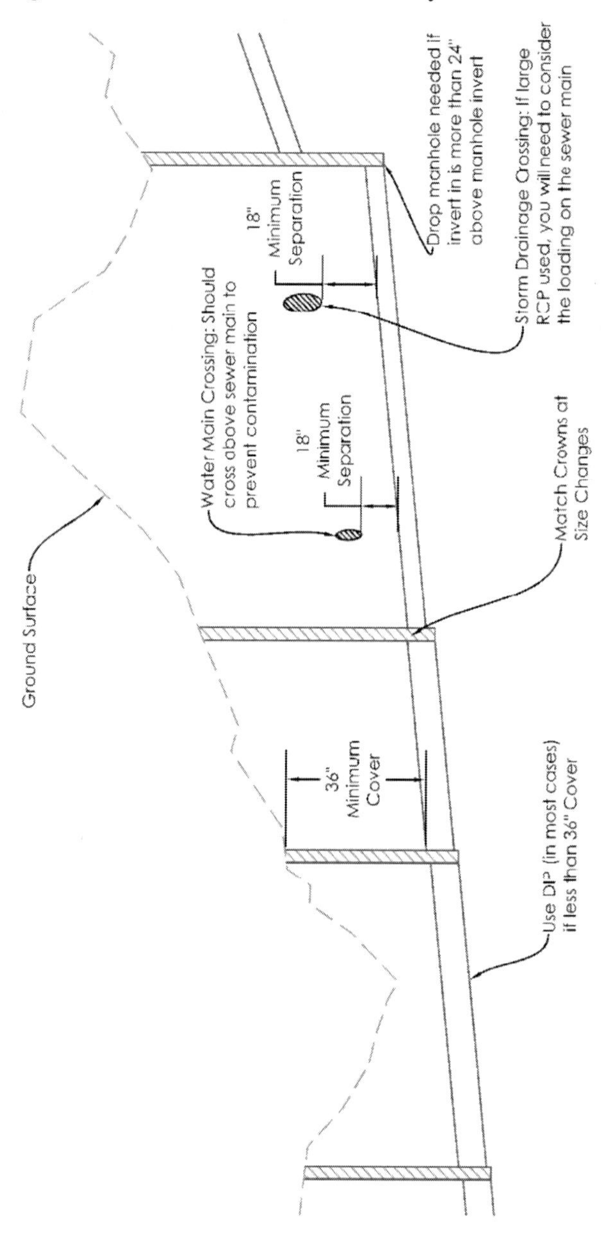

Appendix A
Summary of Formulas

Reynolds Number

$$Re = \frac{VR_d}{v}$$

- *Re = Reynolds number (unitless)*
- *V = Average Velocity (ft/s)*
- R_d *= Hydraulic Diameter (ft)* → $R_d = 4A/P^*$
- *v = kinematic viscosity* (ft^2/s)

Equation of Continuity

$$\rho_1 A_1 V_1 = \rho_2 A_2 V_2$$

$$A_1 V_1 = A_2 V_2 \; or \; Q_1 = Q_2$$

- *p = density*
- *A = Area*
- *V = Velocity*

Energy Equation

$$P_1 + \frac{1}{2}\rho v_1^2 + \rho g h_1 = P_2 + \frac{1}{2}\rho v_2^2 + \rho g h_2$$

$$or$$

$$\frac{P_1}{\gamma} + \frac{v_1^2}{2g} + h_1 = \frac{P_2}{\gamma} + \frac{v_2^2}{2g} + h_2$$

$$P = Pressure\ Energy;$$

$$\frac{1}{2}\rho v^2 = Kinetic\ Energy;$$

$$\rho g h = Potential\ Energy$$

- *P: Pressure*
- *ρ: Density*
- *v: Velocity*
- *h: height*
- *g: gravity*
- *γ: specific gravity (ρg)*

Hydraulic Grade Line

$$HGL = P_1 + \rho g h_1 \text{ , or}$$

$$HGL = \frac{P_1}{\gamma} + h_1 + energy\ added - energy\ lost$$

Energy Grade Line

$$EGL = \frac{P_1}{\gamma} + \frac{v_1^2}{2g} + h_1 + energy\ added - energy\ lost$$

Manning's Equation

Manning's Equation	
Manning's Equation for Flow Velocity (ft/s)	$V = \frac{1.49}{n} R^{\frac{2}{3}} S^{\frac{1}{2}}$
Manning's Equation for Flow Rate (cfs)	$Q = \frac{1.49}{n} R^{\frac{2}{3}} S^{\frac{1}{2}} A$
General Parameters	
n	See Table 2
Radius	$r = \frac{D}{2}$
Hydraulic Radius	$R = \frac{A}{P}$
Angle	$\theta = 2\arccos\left(\frac{r-h}{r}\right)$
Pipe Less than Half Full	
Wetted Perimeter	$P = r\theta$
Depth of Water	$h = y$
Area	$A = \frac{r^2(\theta - sin\theta)}{2}$

Pipe More than Half Full	
Wetted Perimeter	$P = 2\pi r - r\theta$
Depth of Water	$h = 2r - y$
Area	$A = \pi r^2 - \frac{r^2(\theta - sin\theta)}{2}$

Hazen-Williams Equation

$$V = 1.318CR^{0.63}S^{0.54}$$

- *V=Velocity (ft/s)*
- *C= Hazen-Williams Coefficient*
- *R = Hydraulic Radius (ft)*
- *S = Head Loss per Length (ft/ft)*

$$h_f = 10.5L\left(\frac{Q}{C}\right)^{1.85} D^{-4.87}$$

- *h_f = Head Loss*
- *L = Length (ft)*
- *Q = Flow Rate (gpm)*
- *D = Diameter (Inches)*

Minor Losses

$$h_m = \sum k\frac{v^2}{2g}$$

- *k = Minor Loss Coefficient*
- *v = Velocity (ft/s)*
- *g = Gravity (ft/s²)*

Time of Concentration

Sheet Flow

$$T_t = \frac{0.007(nL)^{0.8}}{(P_2)^{0.5}(S)^{0.4}}$$

- *Tt : Travel Time in Hours*
- *n: Manning's n Value (Represents the roughness of the surface)*
- *L: Travel Length in Feet (Not to exceed 100 feet)*
- *P_2: Rainfall for the 2-year, 24-hour storm event in Inches*
- *S: Slope of the surface*

Shallow Concentrated Flow

$$T_t = \frac{L}{3600V}$$

- *L: Flow Length (ft)*
- *V: Average Velocity (ft/s)*

$$V = 16.13(S)^{0.5} \text{ (Unpaved)}$$

$$V = 20.33(S)^{0.5} \text{ (Paved)}$$

Channel Flow

$$V = \frac{1.49(R)^{\frac{2}{3}}(S)^{\frac{1}{2}}}{n}$$

NRCS Runoff Equations

$$Q = \frac{(P - I_a)^2}{(P - I_a) + S}$$

- *Q = Runoff (inches)*
- *P = Rainfall Amount (Inches)*
- *Ia = Initial Abstraction (storage, interception, infiltration, evaporation)*
- *S = Maximum Soil Retention*

$$I_a = 0.2S$$

$$Q = \frac{(P - 0.2S)^2}{(P + 0.8S)}$$

$$S = \frac{1000}{CN} - 10$$

$$CN = \frac{1000}{(10 + 5P + 10Q - 10(Q^2 + 1.25QP)^{.5})}$$

Orifice Discharge

$$Q = Ca\sqrt{2gh}$$

- *Q: Flow Rate (cfs)*
- *a: Orifice Diameter (ft)*
- *g: Gravity (32.2 ft/s2)*
- *h: Height (ft)*

Peaking Factors

$$Peak\ Hourly\ Flow = Peaking\ Factor * \ Average\ Daily\ Flow$$

$$Peak\ Hourly\ Flow = 2.5 * \ 300 = 750\ gpd$$

Flow Monitoring Analysis Results

R-Value

$$R - Value = \frac{RDII\ Volume}{Rainfall\ Volume}$$

Peak Wet Weather Flow Factor

$$Peak\ Factor = \frac{Maximum\ Flow\ during\ Storm\ Event}{Average\ Flow\ during\ Dry\ Period}$$

Pump Station

$$\frac{\Delta V}{\Delta T} = Q_{out} - Q_{in}$$

$$Cycle\ Time\ (T) = TIme\ to\ Fill + Time\ to\ Pump\ Down$$

$$Time\ to\ Fill = \frac{V}{Q_{in}}$$

$$Time\ to\ Pump\ Down = \frac{V}{Q_{out} - Q_{in}}$$

$$T = \frac{V}{Q_{in}} + \frac{V}{Q_{out} - Q_{in}} \rightarrow V = \frac{T}{\left(\frac{1}{Q_{in}} + \frac{1}{Q_{out} - Q_{in}}\right)}$$

Appendix B
Manning's Roughness Coefficients[18]

Material	Manning's n Value
Asbestos cement	0.011
Asphalt	0.016
Brass	0.011
Brick	0.015
Canvas	0.012
Cast-iron, new	0.012
Clay tile	0.014
Concrete - steel forms	0.011
Concrete (Cement) - finished	0.012
Concrete - wooden forms	0.015
Concrete - centrifugally spun	0.013
Copper	0.011
Corrugated metal	0.022
Earth, smooth	0.018
Earth channel - clean	0.022
Earth channel - gravelly	0.025
Earth channel - weedy	0.03
Earth channel - stony, cobbles	0.035

[18] (Engineering Toolbox, 2004)

Material	Manning's n Value
Floodplains - pasture, farmland	0.035
Floodplains - light brush	0.05
Floodplains - heavy brush	0.075
Floodplains - trees	0.15
Galvanized iron	0.016
Glass	0.01
Gravel, firm	0.023
Lead	0.011
Masonry	0.025
Metal - corrugated	0.022
Natural streams - clean and straight	0.03
Natural streams - major rivers	0.035
Natural streams - sluggish with deep pools	0.04
Natural channels, very poor condition	0.06
Plastic	0.009
Polyethylene PE - Corrugated with smooth inner walls	0.009 - 0.015
Polyethylene PE - Corrugated with corrugated inner walls	0.018 - 0.025
Polyvinyl Chloride PVC - with smooth inner walls	0.009 - 0.011
Rubble Masonry	0.017
Steel - Coal-tar enamel	0.01
Steel - smooth	0.012
Steel - New unlined	0.011
Steel - Riveted	0.019
Vitrified Sewer	0.013 - 0.015

Appendix C
Hazen-Williams Coefficients[19]

Material	Hazen-Williams Coefficient (C)
ABS - Acrylonite Butadiene Styrene	130
Aluminum	130 - 150
Asbestos Cement	140
Asphalt Lining	130 - 140
Brass	130 - 140
Brick sewer	90 - 100
Cast-Iron - new unlined (CIP)	130
Cast-Iron 10 years old	107 - 113
Cast-Iron 20 years old	89 - 100
Cast-Iron 30 years old	75 - 90
Cast-Iron 40 years old	64-83
Cast-Iron, asphalt coated	100
Cast-Iron, cement lined	140
Cast-Iron, bituminous lined	140
Cast-Iron, sea-coated	120
Cast-Iron, wrought plain	100
Cement lining	130 - 140

[19] (Engineering Toolbox, 2004)

Material	Hazen-Williams Coefficient (C)
Concrete	100 - 140
Concrete lined, steel forms	140
Concrete lined, wooden forms	120
Concrete, old	100 - 110
Copper	130 - 140
Corrugated Metal	60
Ductile Iron Pipe (DIP)	140
Ductile Iron, cement lined	120
Fiber	140
Fiber Glass Pipe - FRP	150
Galvanized iron	120
Glass	130
Lead	130 - 140
Metal Pipes - Very to extremely smooth	130 - 140
Plastic	130 - 150
Polyethylene, PE, PEH	140
Polyvinyl chloride, PVC, CPVC	150
Smooth Pipes	140
Steel new unlined	140 - 150
Steel, corrugated	60
Steel, welded and seamless	100
Steel, interior riveted, no projecting rivets	110
Steel, projecting girth and horizontal rivets	100
Steel, vitrified, spiral-riveted	90 - 110
Steel, welded and seamless	100
Tin	130
Vitrified Clay	110
Wrought iron, plain	100

Appendix D
Wastewater Loading[20]

Type of Establishment	Hydraulic Loading (GPD)
A. Airport:	
1. Per Employee	8
2. Per Passenger	4
B. Apartments, Condominiums, Patio Homes:	
1. Three (3) Bedrooms (Per Unit)	300
2. Two (2) Bedrooms (Per Unit)	225
3. One (1) Bedroom (Per Unit)	150
C. Assembly Halls: (Per Seat)	4
D. Barber Shop:	
1. Per Employee	8
2. Per Chair	75
E. Bars, Taverns:	
1. Per Employee	8
2. Per Seat, Excluding Restaurant	30
F. Beauty Shop:	
1. Per Employee	8
2. Per Chair	94
G. Boarding House, Dormitory: (Per Resident)	38
H. Bowling Alley:	
1. Per Employee	8
2. Per Lane, No Restaurant, Bar or Lounge	94
I. Camps:	

[20] (South Carolina Department of Health and Environmental Control, 2002)

Type of Establishment	Hydraulic Loading (GPD)
1. Resort, Luxury (Per Person)	75
2. Summer (Per Person)	38
3. Day, with Central Bathhouse (Per Person)	26
4. Travel Trailer (Per Site)	131
J. Car Wash: (Per Car Washed)	56
K. Churches: (Per Seat)	2
L. Clinics, Doctor's Office:	
1. Per Employee	11
2. Per Patient	4
M. Country Club, Fitness Center, Spa: (Per Member)	38
N. Dentist Office:	
1. Per Employee	11
2. Per Chair	6
3. Per Suction Unit; Standard Unit	278
4. Per Suction Unit; Recycling Unit	71
5. Per Suction Unit; Air Generated Unit	0
O. Factories, Industries:	
1. Per Employee	19
2. Per Employee, with Showers	26
3. Per Employee, with Kitchen	30
4. Per Employee, with Showers and Kitchen	34
P. Fairgrounds: (Average Attendance, Per Person)	4
Q. Grocery Stores: (Per Person, No Restaurant or Food Preparation)	19
R. Hospitals:	
1. Per Resident Staff	75
2. Per Bed	150
S. Hotels: (Per Bedroom, No Restaurant)	75
T. Institutions: (Per Resident)	75
U. Laundries: (Self Service, Per Machine)	300
V. Marinas: (Per Slip)	23
W. Mobile Homes: (Per Unit)	225
X. Motels: (Per Unit, No Restaurant)	75
Y. Nursing Homes:	
1. Per Bed	75
2. Per Bed, with Laundry	113

Type of Establishment	Hydraulic Loading (GPD)
Z. Offices, Small Stores, Business, Administration Buildings: (Per Person, No Restaurant)	19
AA. Picnic Parks: (Average Attendance, Per Person)	8
BB. Prison/Jail:	
1. Per Employee	11
2. Per Inmate	94
CC. Residences: (Per House, Unit)	300
DD. Rest Areas, Welcome Centers:	
1. Per Person	4
2. Per Person, with Showers	8
EE. Rest Homes:	
1. Per Bed	75
2. Per Bed, with Laundry	113
FF. Restaurants:	
1. Fast Food Type, Not Twenty-Four (24) Hours (Per Seat)	30
2. Twenty-Four (24) Hour Restaurant (Per Seat)	53
3. Drive-In (Per Car Service Space)	30
4. Vending Machine, Walk-up Deli or Food Preparation (Per Person)	30
GG. Schools, Day Care:	
1. Per Person	8
2. Per Person, with Cafeteria	11
3. Per Person, with Cafeteria, Gym and Showers	15
HH. Service Stations:	
1. Per Employee	8
2. Per Car Served	8
3. Car Wash (Per Car Washed)	56
II. Shopping Centers, Large Department Stores, Malls: (Per Person, No Restaurant)	19
JJ. Stadiums, Coliseums: (Per Seat, No Restaurant)	4
KK. Swimming Pools: (Per Person, with Sewer Facilities and Showers)	8
LL. Theaters: Indoor (Per Seat), Drive In (Per Stall)	4

References

Dewberry. (2008). *Land Development Handbook: Planning, Engineering, and Surveying* (3rd ed.). New York: McGraw-Hill.

Engineering Toolbox. (2004). *Hazen-Williams Coefficients*. Retrieved June 20, 2018, from https://www.engineeringtoolbox.com/hazen-williams-coefficients-d_798.html

Engineering Toolbox. (2004). *Manning's Roughness Coefficients*. Retrieved June 20, 2018, from https://engineeringtoolbox.com/mannings-roughness-d_799.html

Engineering Toolbox. (2004). *Minor Loss Coefficients in Pipes and Tube Components*. Retrieved June 20, 2018, from https://www.engineeringtoolbox.com/minor-loss-coefficients-pipes-d_626.html

Greater Lakes - Upper Mississipi RIver Board of State and Provincial Public Health and Environmental Managers. (2014). *Recommended Standards for Wastewater Facilities*. Health Research, Inc., Health Education Services Division. Retrieved December 29, 2018, from Recommended Standards for Wastewater Facilities: http://www.health.state.mn.us/divs/eh/water/tenstates/documents/wstewtrstnds2014secured.pdf

Metcalf & Eddy, Inc. (1981). *Wastewater Engineering: Collection and Pumping of Wastewater*. New York: McGraw-Hill.

Randolph, J. (2004). *Environmental Land Use Planning and Management*. Washington: Island Press.

South Carolina Department of Health and Environmental Control. (2002). R.61-67 Standards for Wastewater Facility Construction.

United States Department of Agriculture. (1986, June). Urban Hydrology for Small Watersheds: Technical Release 55.

Made in United States
North Haven, CT
10 March 2023